photo: Getty Images

Discovery EDUCATION | MATH TECHBOOK

ALGEBRA 1 WITH HONORS

TEACHER GUIDE FOR FLORIDA INTERACTIVE STUDENT EDITION

Log in to Math Techbook at
www.DiscoveryEducation.com

ISBN 13: 978-1-68220-486-3

Library of Congress Cataloging-in-Publication Data available upon request from the Library of Congress.

Printed in the United States of America.

1 2 3 4 5 6 7 8 9 10

Discovery EDUCATION™

800-323-9084
One Discovery Place, Silver Spring, Maryland 20910

Table of Contents

ANSWER KEYS

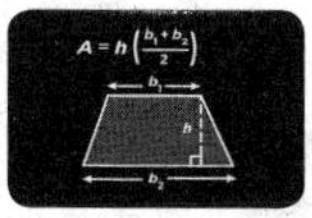

UNIT 1: Foundations of Algebra

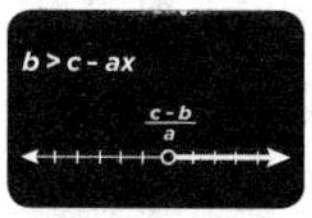

UNIT 2: Equations and Inequalities

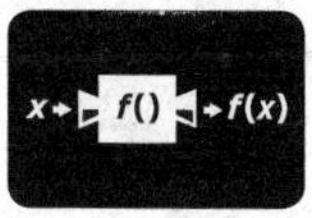

UNIT 3: Functions

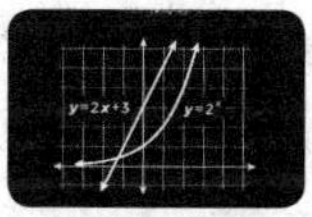

UNIT 4: Graphs of Functions

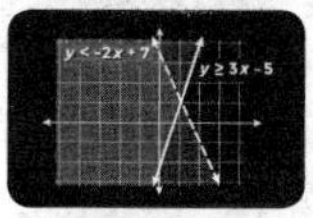

UNIT 5: Systems of Equations and Inequalities

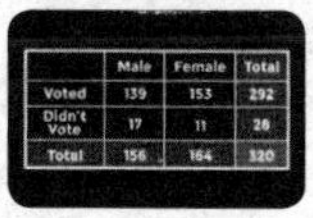

UNIT 6: Descriptive Statistics

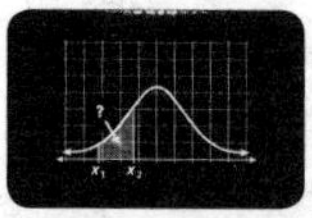

UNIT 7: Data Modeling

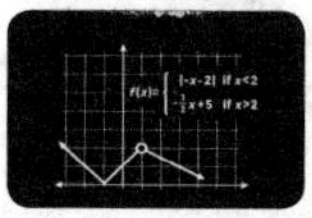

UNIT 8: Nonlinear Functions

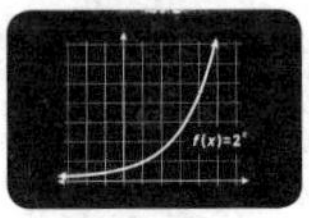

UNIT 9: Exponential Functions

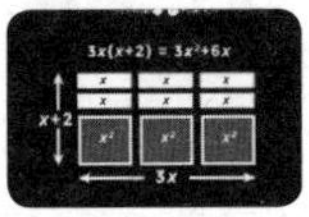

UNIT 10: Polynomials

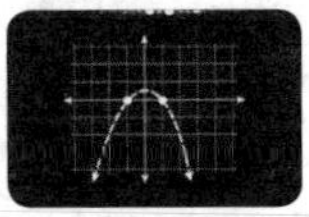

UNIT 11: Quadratic Expressions and Equations

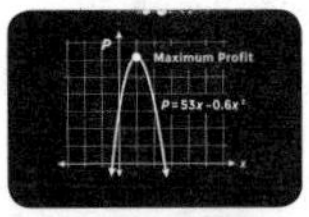

UNIT 12: Graphs of Quadratic Functions

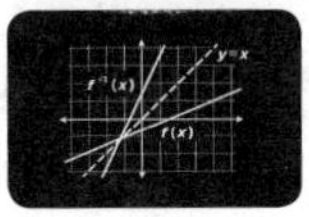

UNIT 13: Recursive, Explicit, and Inverse Functions

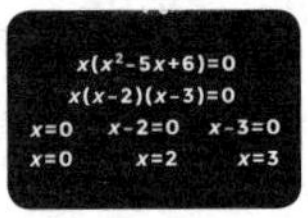

UNIT 14: Polynomial Expressions and Equations

UNIT 15: Rational Expressions and Equations

Letter to the Teacher

Dear Teacher,

Welcome to Math Techbook™! This Teacher Guide provides an overview of Discovery Education Math Techbook and contains answers to questions that appear in the *Interactive Student Edition*. This guide will help you explore the instructional techniques and conceptual framework of Math Techbook, as well as prepare you for your key role as a facilitator of mathematics learning.

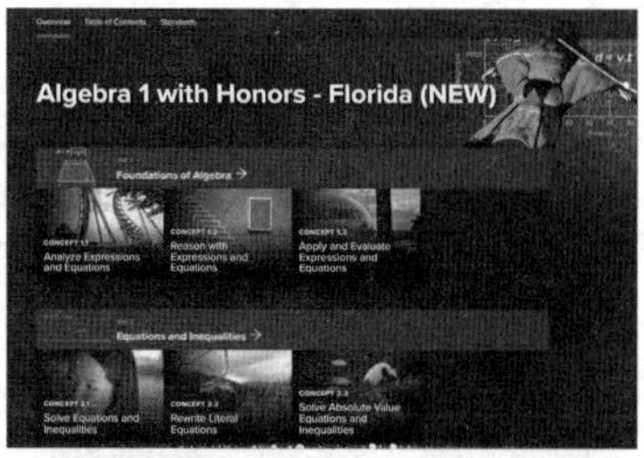

In this Teacher Guide, you'll find two major sections:

- OVERVIEW: Math Techbook is different from other programs you've used, and this section will help you understand how and why. In addition, you'll learn how to navigate Math Techbook and effectively use the Discover-Practice-Apply instructional cycle.
- ANSWER KEYS: The *Interactive Student Edition* contains more than 500 questions, and answers for each question appear in this Teacher Guide. Links are also provided to solutions for the Apply problems, which are typically more complex multistep problems.

The *Interactive Student Edition* is not the primary instructional resource. Rather, it is a print subset of Math Techbook, a fully digital, core instructional resource. Math Techbook engages students with dynamic content designed to help students develop deep mathematical understanding and apply it to the world around them. As the classroom facilitator, you'll be equipped with a comprehensive resource that includes student activities, model lessons, and a dashboard that shows student progress—all in one, easy-to-navigate site.

Math Techbook provides the all-important learning opportunities that encourage students to develop conceptual understanding, procedural fluency, and application in a structured and balanced manner. It offers many digital resources, none of which would be available in a print curriculum or PDF-driven sites:

- MATH TOOLS: A suite of digital tools, including a graphing calculator, white board, and dynamic geometry software, are included with Math Techbook.
- INTERACTIVES: Within Math Techbook, students are exposed to online activities that allow the user to interact with them, playing games that involve everything from skeleton armies and absolute value in Grade 6 to caffeine elimination and exponential decay in Algebra II.
- VIDEOS: Because you expect nothing less from Discovery Education, Math Techbook includes a vast assortment of videos that engage students in new topics, present a problem situation, or explain concepts.
- TECHNOLOGY-ENHANCED ITEMS (TEIs): These computer-delivered questions allow special interactions for collecting data, going far beyond the typical multiple-choice and constructed-response items of the past. All data gathered by the TEIs is reported in the dashboard.
- DASHBOARD: Students, teachers, and parents can monitor student progress using real-time data.

What Makes the ISE Interactive?

As previously mentioned, the *Interactive Student Edition* is not a complete instructional tool. Rather, it is designed for students to interact with the comprehensive digital Math Techbook. Quick-response (QR) codes are provided throughout the resource to direct students to specific learning resources.

To use the QR codes, you will need a QR reader. Readers are available for phones, tablets, laptops, desktops, and virtually any device in between. Download a free QR reader in the App Store or Google Play. To access Math Techbook resources, follow these steps:

1. Open the QR code reader on your device.
2. Hold your device so the QR code is visible within your device's screen. One of two things will happen:
    - The device may automatically scan the code; or,
    - The device will scan the code when you press a button, similar to taking a picture.
3. Once scanned, the QR code will direct you to a page or resource on the web.
4. For resources in Math Techbook, you may need to log in with your username and password the first time you access a QR code. After that, you won't need to log in again, unless you sign out or remain inactive for too long.

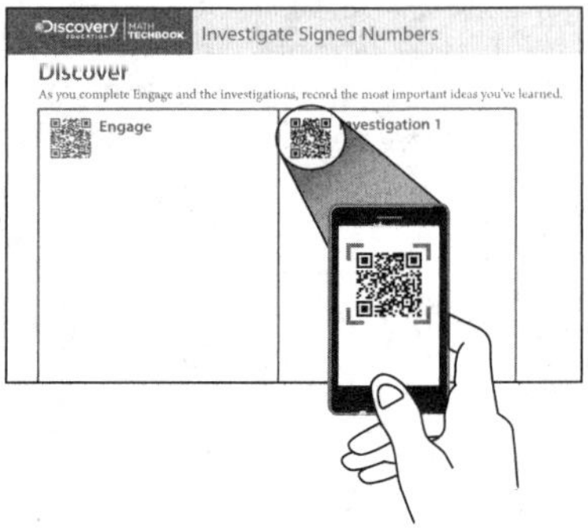

Scan this QR code to access a video that provides a deeper introduction to Math Techbook:

As you implement Math Techbook and its helpful digital content, prepare for a transformation! Enjoy the journey, and best of luck for a fantastic year!

Sincerely,

The Discovery Education Math Team

How to Use Math Techbook™

The Techbook Difference

Discovery Education Math Techbook is a dynamic, digital resource that empowers students to take ownership of their learning. Compatible with any hardware used by school districts, Discovery Education Math Techbook elevates student achievement through a combination of three critical elements.

I. **Content:** Exclusive content from Discovery Education engages students and brings math to life, while meeting all the demands of the standards.

II. **Instruction:** Research-based, inquiry-driven instructional design balances conceptual understanding, procedural fluency, and application to relevant problems.

III. **Technology:** Powerful, simple-to-use technology saves teachers time and eases the transition to digital resources.

It is the convergence of all three elements that makes Math Techbook truly unique. Math Techbook is a powerful resource that both inspires today's digital native students and brings transformational change to the classroom. The result is a marked increase in student engagement and learning.

The Power of Three

Navigating Math Techbook

After signing in and navigating to your desired Math Techbook course, **select a unit** you would like to teach. The units in each grade level provide in-depth coverage of the standards.

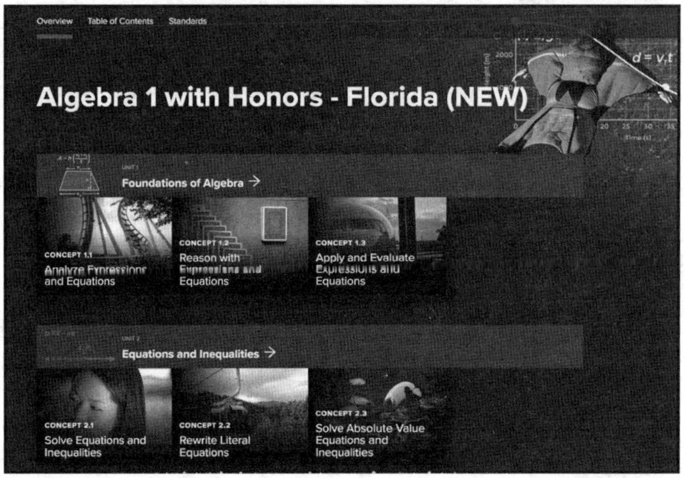

Within each unit are one or more concepts to explore. Each unit also includes a Unit Opener that presents a real-world situation related to the unit's focus. **Select a concept** by clicking on a concept title or an image.

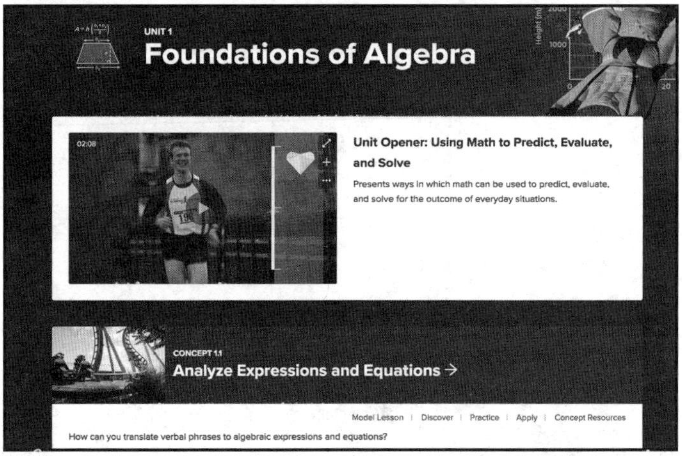

Navigate using the Discover, Practice, Apply, and Model Lesson tabs below the title to **explore the components** of a concept. The Model Lesson provides all of the information you need to successfully teach each concept: step-by-step suggestions for using the interactives, questions for discussion, and assessment items that will assist your students in the exploration and deep learning of a concept.

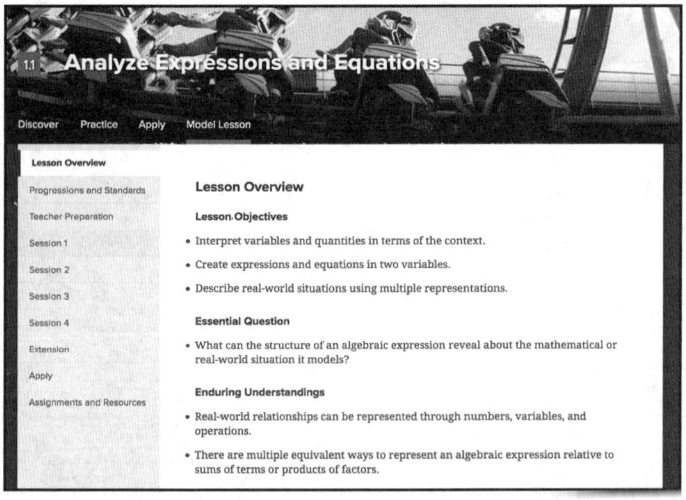

Within Discover, **explore the content** that your students will engage with. Through the activities in Engage and Investigate, students will be presented with real-world problems. They will use interactives and digital tools to make sense of problems and find solutions.

After completing the investigations, students will practice their new skills in a game-like environment and apply their skills in a way that requires more rigor and deeper connections. You can view exercises for procedural fluency in Practice and view real-world problems worth solving in Apply.

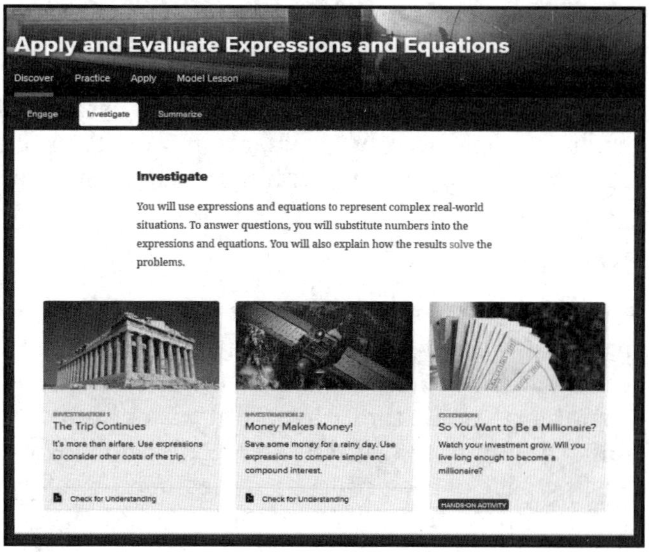

Discovery Education's Math Techbook also includes tools that assist students' learning and help you keep track of each student. The **Teacher Learning Dashboard** shows you student results. Students also have a version of the Dashboard where they can review their work.

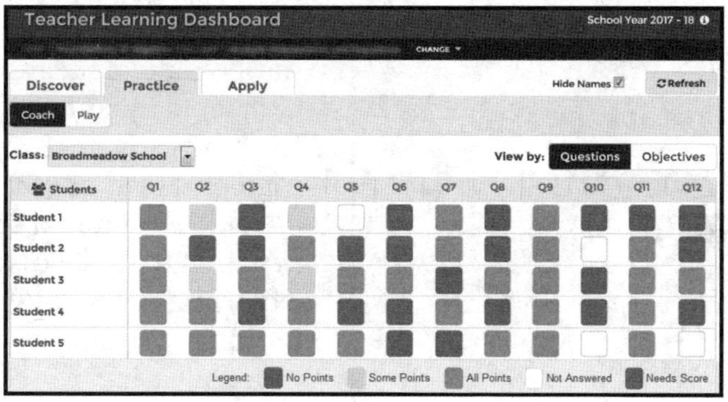

Math Tools include a graphing calculator, dynamic geometry software, and other tools your students can use to explore math interactively. Many of these tools are embedded within lessons, but you can always choose to use them in your own way and have students share their work with you or the entire class.

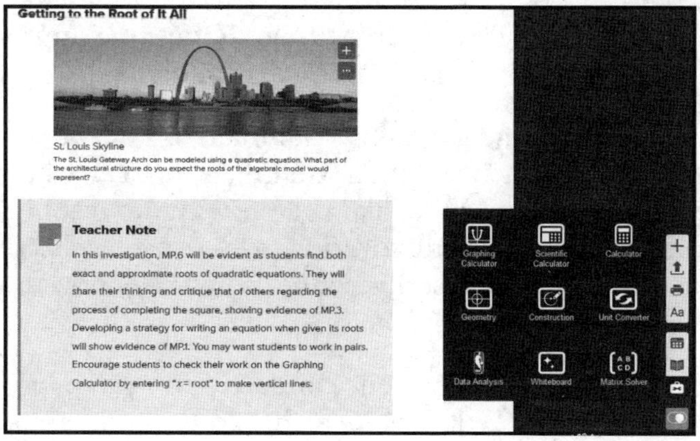

Discover – Practice – Apply

Discovery's Math Techbook is based on a Discover, Practice, Apply instructional cycle. This cycle of learning enables students to solve problems in more than one way using more than one modality, allowing students to reach mastery learning with more depth. Students retain more of what they have learned.

Discover

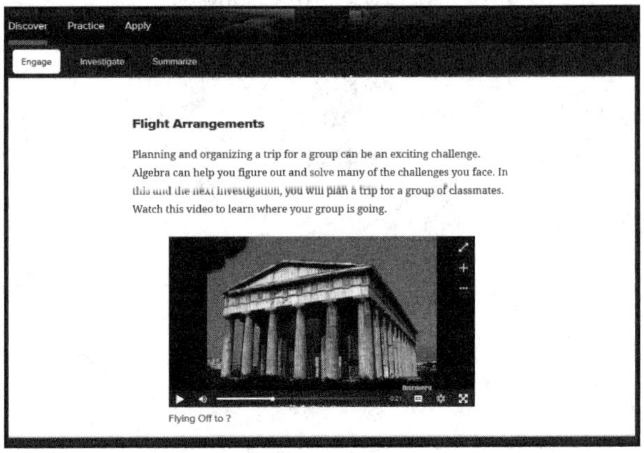

Engage: An intriguing video or thought-provoking image and situation captures students' attention and ignites their curiosity. This activity motivates students to learn while providing an opportunity to activate prior knowledge, connecting back to concepts and terms previously learned. The first page of each concept in the *Interactive Student Edition* introduces students to the skills and topics they will encounter. Each concept includes one or more Essential Questions that will help students focus on the key ideas. Students may return to the Essential Questions as a form of self-assessment at the end of each concept.

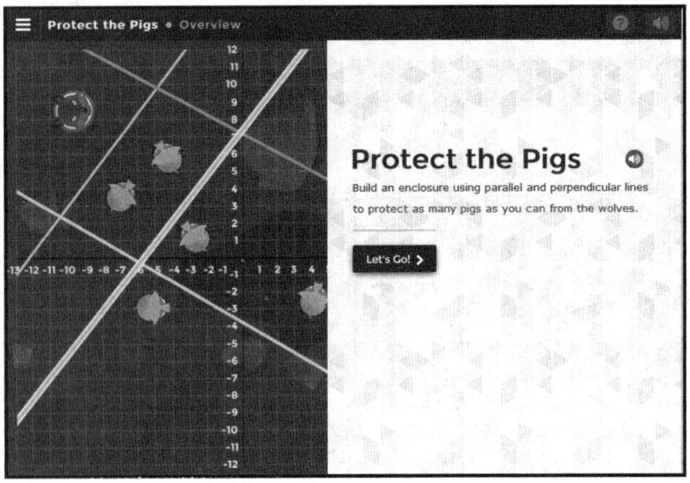

Investigate: This section is the main component of the concept. Investigations keep students actively engaged, thinking, and developing mathematical meaning in everyday settings. Concepts include two to six investigations. Each investigation immerses students in mathematical or real-world scenarios. They provide the guided support necessary for students to make discoveries by conducting experiments, making observations, testing ideas, looking for patterns, and summarizing results. Using a variety of recommended instructional and classroom management strategies, students deepen their understanding by sharing their discoveries, listening to the conjectures of their classmates, and critiquing arguments.

Investigations scaffold the development of a concept with activities that approach the concept in different ways. The activities within any investigation may include the following activity types:

- **Interactives** are exploratory activities and games that put math in students' hands and allow them to explore a concept in depth. Some interactives take a scientific approach, where students create several test cases and compare results. Others feel more like a game, with motivational elements that encourage strategic, critical thinking. Students learn by discovering math concepts that lead to successful outcomes. As a follow-up, students analyze patterns, describe connections they noticed, and summarize what they learned in a formal way.

- **Hands-On Activities** allow students to experience and discover mathematical relationships in ways similar to the interactives, but they do not require technology. Students can work independently or collaboratively in small groups.

- **Technology-Enhanced Items** encourage students to summarize their thinking and allow them to become familiar with new question formats they will see on high-stakes assessments.

In the *Interactive Student Edition*, students can record the results of their learning from within Engage and Investigate in their own words. Doing so may bring up further questions, giving you the opportunity to address any misconceptions and help students solidify their learning.

To give students practice and to confirm student understanding, the *Interactive Student Edition* includes a Check for Understanding for each investigation. Used as a nightly assignment, Check for Understanding assesses the depth of knowledge your students retained from the day's learning.

The *Interactive Student Edition* also includes a summary of key concepts and skills, a condensed version of what is available at the end of each concept. When students have a partial understanding of the concept and require reinforcement, the Summary section gives students the opportunity to confirm what they learned during the investigations and address misconceptions that may return when students are working independently.

Practice

The Practice section consists of two parts, Coach and Play, each containing a collection of exercises for your students to sharpen their skills.

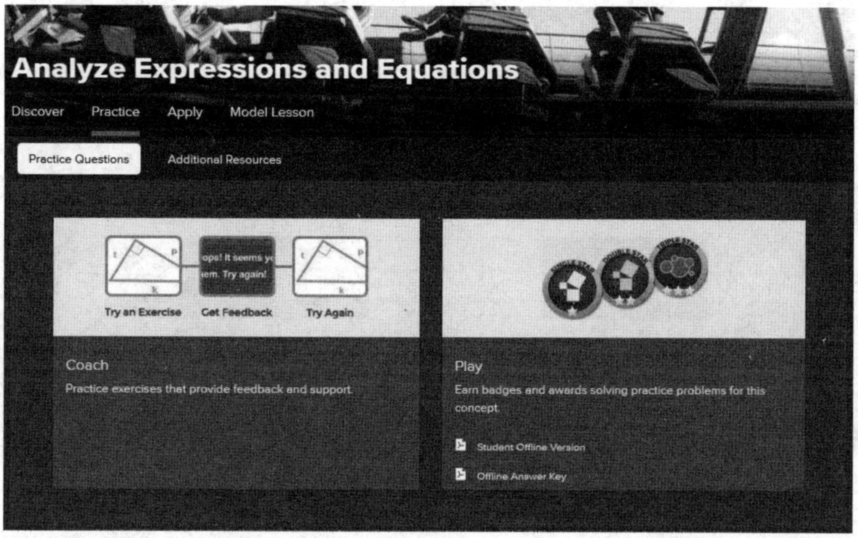

The items within **Coach** provide a set of guided practice exercises. Immediate, specific feedback in the form of tips or suggestions directly related to a student's error is provided when an incorrect answer is entered. Feedback often directly addresses the student's misconception and always guides him or her to the correct solution path. If a student is unable to identify the correct answer in three attempts, the correct answer is provided with instruction. This allows students to continue to learn as they answer each question. Similar to athletics, Coach provides necessary support, giving suggestions, advice, just-in-time instruction, and a virtual high-five for student success.

The items within **Play** provide a set of independent practice exercises. Just as in athletics, the coach cannot step in and provide assistance during the big game. The player's success is up to his or her own strength and abilities. In Play, students answer questions independently to earn badges for their knowledge. Students earn a star badge for the concept after successfully completing five problems, and they get more stars added to their badge as they correctly solve more problems. Students are motivated to continue solving problems to improve their score and earn more stars.

The *Interactive Student Edition* contains a set of Practice Exercises, and a printable version of the Play items is available online through the Student Offline Version button. Consequently, there are two options for students to practice their skills when they're away from the computer.

Apply

Apply culminates the learning cycle by ensuring that students not only understand the concept, but can also use what they've learned in their everyday lives. The problems in Apply go beyond traditional word problems. Just as in real-world situations, all the information that students need may not be provided. Students are often required to conduct research to complete the problem, determine what information they need, and ignore information they do not need. Apply provides real-world scenarios that involve multiple steps and reasoning, and students must submit models, equations, drawings, or explanations to support their solutions. Most importantly, Apply situations often have more than one feasible solution, depending on the model that students use and the data they collect. Math Techbook provides teacher support for every Apply scenario by including a rubric with 4-click scoring of student solutions submitted online. Apply can be used as homework, formative or summative assessment, or for project-based assessment.

One Apply problem is included in the *Interactive Student Edition*, making it easy to use for offline assignments.

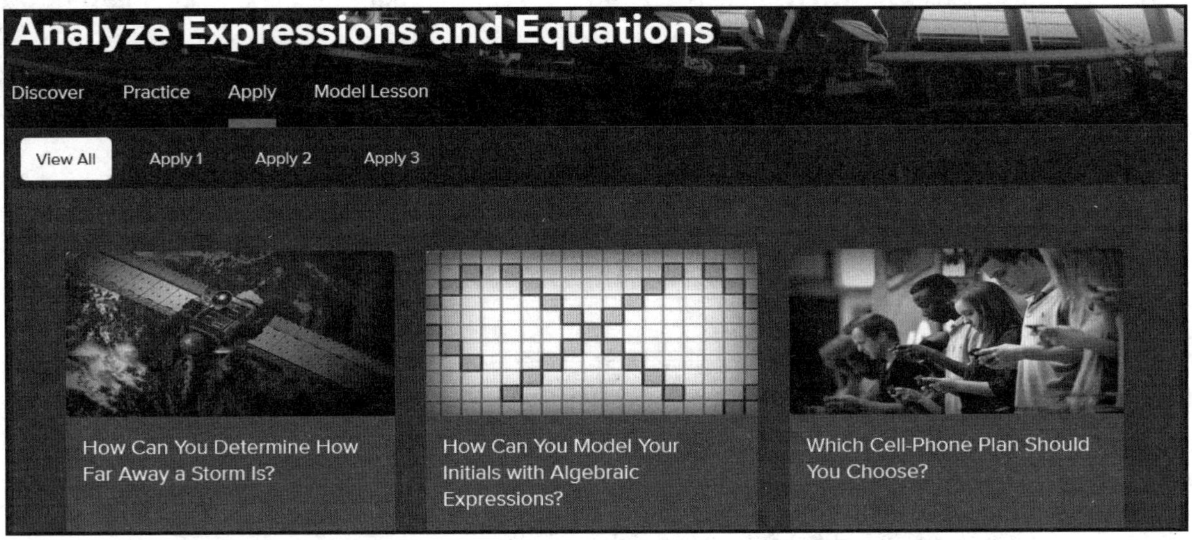

Discovery EDUCATION™ | MATH TECHBOOK™

Answer Keys for Check for Understanding and Practice Exercises

1.1 Analyze Expressions and Equations

Check for Understanding: Investigation 1

1. B, C
2. A, D
3. 15,550 cm^2
4. Sample response: In Brian's expression the height of the box is $H = L + 50$ so $H - 50 = L$. Also the width of the box is $W = L - 20$ so I can write the width as $W = (H - 50) - 20 = H - 70$. The expression for surface area is then $2(H - 50)(H - 70) + 2H(H - 70) + 2H(H - 50)$.

Check for Understanding: Investigation 2

1.

x	1	2	3	4	5	6	7	8	9	10
value	30	27	24	21	18	15	12	9	6	3

2. D
3.

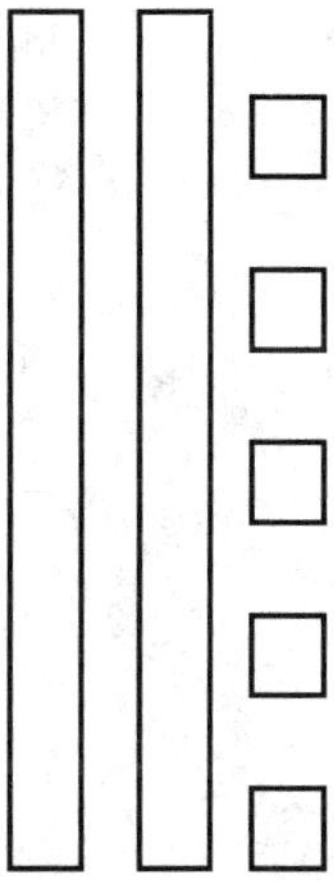

4. A table is useful for identifying patterns and trends, and an expression is useful for finding values that far exceed what's in the table. Tiles allow for quick visual comparisons. Each type of representation provides different information, and comparing the various representations often provides insight about the situation.

Check for Understanding: Investigation 3

1. Annike: $11n$, Celia: $14(n - 1)$ or $14n - 14$ or equivalent
2. A
3. 11:40 a.m.
4. I set the equations from Question 1 equal to each other and then solved. Which gave $n = 4\frac{2}{3}$, so I added that to 7 a.m. For another method, I know Annike had already gone 11 miles by 8 a.m. Celia bikes 3 more miles every hour than Annike, so it would take her $11 \div 3 = 3\frac{2}{3}$ hours to catch Annike. That would also give an answer of 11:40 a.m. I also could have graphed both of those equations, and the intersection would give the time they had gone the same distance.

Analyze Expressions and Equations *(continued)*

Practice Exercises

1. C
2. A, D, F
3. B
4. B
5. A, D, E
6. A
7. B, G
8.

Number of Flower Pots, n	1	2	3	4	5
Height of Stack in cm	28	34	40	46	52

$6n + 22$

9. $82 - 3h$
10. $y = \$81$

Apply

How Can You Model Your Initials with Algebraic Expressions?

This problem allows students to determine a pattern in the terms of a sequence and write an algebraic expression to find any term in the sequence based on the term number. Starting with the first letter of their name, students create 3 expanding variations of the letter, applying an arithmetic change to the number of boxes. Students will create an algebraic expression to represent this expansion and use their algebraic expression to determine the number of squares needed to make the 150th letter in the sequence they created. Students may use the online grid or complete the pattern on grid paper.

The Apply problems are rich performance tasks that may have a variety of open-ended solutions. One sample student response for this problem can be found in the Model Lesson.

1.2 Reason with Expressions and Equations

Check for Understanding: Investigation 1

1. $80n + 150$ (or equivalent)
2. $95n$
3. \$2,070, \$2,280
4. A
5. Evaluating one of the expressions for total cost requires performing one or two operations, whereas another method such as performing repeated addition could require performing an operation 20 times. Therefore, evaluating an expression saves time and provides less opportunity for making a mistake in calculating total cost.

Check for Understanding: Investigation 2

1. B
2. 43.225 m or rounded values
3. Sample response: As the height increases, the denominator gets bigger so the fraction will be smaller and the length decreases. For example, suppose the volume is $100\,\text{cm}^3$, the width is 2 cm and the height is 5 cm, then the length is $L = \frac{100\,\text{cm}^3}{(2\,\text{cm})(5\,\text{cm})} = 10\,\text{cm}$. If the height increases to 20 cm, then the length becomes $L = \frac{100\,\text{cm}^3}{(2\,\text{cm})(20\,\text{cm})} = 2.5\,\text{cm}$.

Check for Understanding: Investigation 3

1. A. III.
   B. I.
   C. IV.
   D. II.
2. A, B
3. \$3.87
4. C

Check for Understanding: Investigation 4

1. Each value is 2 more than n^2
2. 2,211
3. C
4. 2, 6, 240, $n^2 + n$

Reason with Expressions and Equations *(continued)*

Practice Exercises

1. C
2. C
3. $87.45x + 69.96$
4. D
5. A, D
6. D
   B, D
   B, D, E
   B, D, E
7. B
8. C
9. D
10. $P = n - (0.3n)$

Apply

How Much Should You Charge to Make the Most Profit?

In this problem, students work in pairs or small groups to choose an item to sell for a fundraiser at a street fair. Students research costs and develop algebraic expressions to describe the expenses and profits related to the item they are selling. Students analyze the costs and profits to determine which item to sell in order to make the most profit.

Students will include all citations for all Internet and other sources they used to conduct their research and document the data they used to solve this Apply problem.

Sample data has been provided for students who have difficulty in completing the research. See the Hands-On Activity: How Much Should You Charge to Make the Most Profit?

The Apply problems are rich performance tasks that may have a variety of open-ended solutions. One sample student response for this problem can be found in the Model Lesson.

1.3 Apply and Evaluate Expressions and Equations

Check for Understanding: Investigation 1

1. A. $E = 50t + 100$

   B. $D = 10\left(\frac{s}{100}\right) + 100$

   C. $R = 10p + 50\left(\frac{p}{50}\right)$

2. C

3. Area = 3,750 sq. ft
   Cost equation: $D = 10(37.5) + 100$
   Cost: $475.00
   Total Decorating Cost: $675.00

   Area = 1,000 sq. ft
   Cost equation: $D = 10(10) + 100$
   Cost: $200.00

4. $R = 10(500) + 50(10)$,
   $5,500.00

5. Total expenses are $6,675.00. 500 students are expected to attend. To make at least $26,675 (20,000 + 6,675), the committee should charge $55.00 ($53.35) per ticket.

Check for Understanding: Investigation 2

1.

Year	0	1	2	3
Account Balance USD	20,000	21,000	22,000	23,000

2.

Year	0	1	2	3
Interest Earned	0	1,000	1,050	1,102.50
Account Balance USD	20,000	21,000	22,050	23,102.50

3. A

4. A

5. The equation $A = P(1 + rt)$ calculates simple interest. The equation $A = P(1 + r)^t$ calculates compound interest. When the amount of money grows linearly and at a constant rate, I know it is based on the linear equation that calculates simple interest. When the amount of money grows at a much greater rate, I know it is based on the exponential growth represented by the equation that calculates compound interest.

Practice Exercises

1. C
2. B, E
3. B
4. A, B, C, D
5. 50, 8, 4
6. $6(s-4)$, 12, 8
7. $18 + nf$
8. B, C, F
9. Equation: F, G, E
   Width: H
   Length: J
10. $1,035.57

Apply and Evaluate Expressions and Equations *(continued)*

Apply

How Much Will It Cost to Drive across the Country?

In this problem, students conduct research to gather costs related to taking a road trip across the country. They define variables and write algebraic expressions for the cost of gasoline, the number of days the trip will take, and additional costs for meals and lodging. The expressions are evaluated using the costs they collected during their research and then combined to determine an estimated total cost for the trip.

Students will include all citations for all Internet and other sources they used to conduct their research and document the data they used to solve this applied problem.

The Apply problems are rich performance tasks that may have a variety of open-ended solutions. One sample student response for this problem can be found in the Model Lesson.

2.1 Solve Equations and Inequalities

Check for Understanding: Investigation 1

1. step 1: B (or A)
step 2: D (or C)
step 3: A (or B)

2. $2x - 10$, $56 + 6x - 30$

3. A. $x = 19$
B. $x = -4$

4. First, Lucy combined the common terms using subtraction on the left side of the equation to get $5(3x - 2) = 15 + 10(3x - 2)$. Then, Lucy used the division property of equality to divide both sides of the equation by 5: $\frac{5(3x-2)}{5} = \frac{15 + 10(3x-2)}{5}$. After dividing by 5, the equation reduces to $(3x - 2) = 3 + 2(3x - 2)$.

Check for Understanding: Investigation 2

1. A, D, E

2. no solution, one solution, infinitely many solutions, one solution

3. $x = 0$

4. $b = 5$

5. Combining terms by adding or subtracting fraction coefficients is more complicated because the fractions must be rewritten with a common denominator. Janae can use the multiplication property of equality to multiply the equation by 16 (or any multiple of 16) so that all the denominators will cancel. All coefficients are now integers and the terms combine more easily.

Check for Understanding: Investigation 3

1. C, D, F

2. A. II.
B. V.
C. I.
D. III.
E. VI.
F. IV.

3. D

4. $g \leq 5.7$ gallons

5. Sammy is correct and the direction of the inequality should not change. To solve, divide by positive 3. The inequality only changes direction if multiplying or dividing by a negative value. The sign of the result doesn't matter.

Solve Equations and Inequalities *(continued)*

Check for Understanding: Investigation 4

1. B
2. A
3. 16 in.
4. 3.5, 3
5. Both occur when the variable terms reduce to a value of zero. A contradiction occurs when the reduced inequality is always false: $2x - 4 > 2x + 6$ reduces to $-4 > 6$, which is false. An identity occurs when the reduced inequality is always true: $2x + 5 > 2x + 3$ reduces to $5 > 3$, which is always true.

Check for Understanding: Investigation 5

1. Smart Storage

Months	Cost in $ at Stack 'n Store	Cost in $ at Smart Storage
1	325	185
2	475	370
3	625	555
4	775	740
5	925	925
6	1075	1110

2. Five months
3. Stack 'n Store: $C = 150m + 175$, Smart Storage: $C = 185m$
4. A
5. To make a profit, the revenue must be greater than the cost, or $R > C$. Write the inequality $10x > 4x + 150$ and solve:
   $10x > 4x + 150$
   $6x > 150$
   $x > 25$

Practice Exercises

1. A. 14
   B. −2
2. E, A, D, B, F
3. −5
4. C, D
5. $t < 87$
6. B
7. No solution: B, F
   Exactly one solution: A, D
   Infinitely many solutions: C, E
8. −13, $-2x$
   Note: The first answer can be any value less than −12.
9. B
10. C

Solve Equations and Inequalities *(continued)*

Apply

How Should a Vendor Stock an Ice Cream Cart?

This problem provides students with an opportunity to show that they can model real-world constraints by writing linear inequalities in one variable and that they can identify values of the variable that satisfy a linear inequality. Note that this problem also requires students to write a linear inequality in two variables, to graph a system of three inequalities on a coordinate plane, and to identify points in the bounded solution region of the system. Because much of this content is not covered in the investigations of this concept, this problem is best suited as an extension for more-advanced students. Note that systems of linear inequalities will be covered in detail in Unit 5.

- Before students begin work on this problem, you may want to provide them with an example of a linear inequality in two variables, such as $x + y \leq 4$. Show students how to graph the inequality and how to identify ordered pairs that represent solutions. Alternately, you can have students research linear inequalities in two variables on their own.
- One way to alter this problem to make it suitable for all students would be to omit the criteria related to graphing (criteria 2 and 3). Students can write inequalities to model the constraints and then develop their own methods of determining five different ways to stock the cart. Students' methods may include guess and check or making a table of values.

The Apply problems are rich performance tasks that may have a variety of open-ended solutions. One sample student response for this problem can be found in the Model Lesson.

2.2 Rewrite Literal Equations

Check for Understanding: Investigation 1

1. D
2. 10.5 feet
3. $4T - 196$ or equivalent
4. A, D
5. This equation is not a literal equation. A literal equation has at least two variables. This equation has only one variable, r. The value π is a constant, not a variable.

Check for Understanding: Investigation 2

1. 10.2 cm and 30.5 cm
2. 2,262 cm^3
3. C
4. $\sqrt{\frac{V}{\pi h}}$, 8.8 cm
5. The diameter is twice the radius. If the diameter and height are the same, then the height must be twice the radius, or $h = 2r$. Substituting $2r$ for h in the volume formula gives $V = \pi r^2(2r)$ or $V = 2\pi r^3$. Solving the formula for r,

$$V = 2\pi r^3$$

$$\frac{V}{2\pi} = r^3$$

$$r = \sqrt[3]{\frac{V}{2\pi}}$$

The only variable used to calculate r is V, which is the volume.

Practice Exercises

1. A
2. C, F
3. C
4. D
5. B
6. $dv, \frac{m}{D}$
7. B, D
8. C
9. RI, 12
10. B, D, G

Rewrite Literal Equations *(continued)*

Apply

What Problem Can I Solve with a Formula?

This problem allows students to apply their understanding of literal equations in the context of solving a real-world problem. It also provides students with the opportunity to conduct self-directed research about a literal equation or formula of their choice.

- If students will not have access to the Internet as they work on this problem, they can use a literal equation with which they are already familiar, such as the surface area formula for a rectangular prism or the area formula for a trapezoid. You may wish to specify that students avoid using literal equations that were covered previously in this concept.
- Some students may not be able to solve for every variable in the literal equation they choose. For example, most students at this level will not yet possess the algebraic skills to solve the equation $h = -16t^2 + vt$ for the variable t. If students find that they cannot solve for one of the variables in their literal equation, suggest that they choose a different variable to solve. For the previously cited example, students could solve for v instead of t.

The Apply problems are rich performance tasks that may have a variety of open-ended solutions. One sample student response for this problem can be found in the Model Lesson.

2.3 Solve Absolute Value Equations and Inequalities

Check for Understanding: Investigation 1

1. B, C, F
2. D
3. C, D, E, F
4. $-4 < x \leq 1$, or $x > -4$ AND $x \leq 1$, or $(-4, 1]$, or equivalent

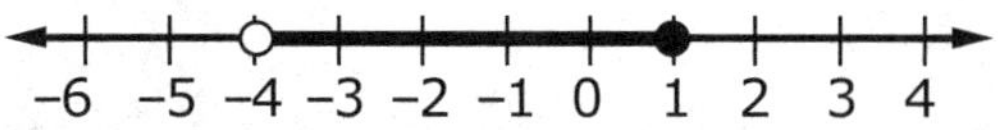

5. This compound inequality is an intersection of two simple inequalities both involving the variable x. They are joined by the word "and." Ericka's solution states that the value of x must be less than −5 and the value of x must be greater than −2. There are no numbers for which both statements are true.

Check for Understanding: Investigation 2

1. B
2. A. IV. B. I. C. III. D. II. E. IV.
3. D
4. 0.5, 1.5
5. Both intervals contain all real numbers between −4 and 12 in the solutions set because these values are used as the endpoints. However, the parentheses on Steve's interval indicate that the values −4 and 12 are not part of the solution set. On the other hand, the brackets on Lynn's interval indicate that the values −4 and 12 are part of the solution set.

Check for Understanding: Investigation 3

1. $3n$, $1.50n + 10$, 30
2. no pass for $15
3. 7, 13
4. A. II. B. VII. C. IV. D. VI. E. III.
5. Jodie is not correct. The simple inequalities are joined by the word OR so Jodie should form a disjunction with the solution sets. Her first inequality is a contradiction, so it has no solution. However, the second inequality, $y \leq 8$, does have solutions. Her final solution to the compound inequality is the union of the two solution sets, or $y \leq 8$.

Check for Understanding: Investigation 4

1. 8, 5, 5, 8, −3, 13 (The answers in the last two blanks are interchangeable.)
2. A. IV. B. I. C. V. D. III.
3. A, D
4. $x = -14$ OR $x = 18$, in any order
5. First, find the point that is the same distance from −1 and 5. This midpoint between −1 and 5 is 2. The distance from 2 to −1 is 3 units. The distance from 2 to 5 is 3 units. So, an equation with the solutions −1 and 5 is $|x - 2| = 3$. (Any positive multiple of this equation would be acceptable as well.)

Solve Absolute Value Equations and Inequalities *(continued)*

Check for Understanding: Investigation 5

1. $|p - 120| = 4$ or $|120 - p| = 4$, 124, 116
2. one solution, two solutions, one solution, no solutions, two solutions
3. 1, −4 (either order correct)
4. D
5. There are two solutions when $|x - 8|$ is equal to a positive number. This means that $b < c$. There is one solution when $|x - 8|$ equals 0. This means that $b = c$. There is no solution when $|x - 8|$ is equal to a negative number. This means that $b > c$.

Check for Understanding: Investigation 6

1. C
2. C, F, J
3. D
4. b can be any integer such that $b \geq 10$
5. The solution set is a single interval when the absolute value of an expression includes "less than" a positive number because $|x| < a$ is a conjunction. The solution set lies to the left and to the right of a single interval when the absolute value of an expression includes "greater than" a positive number because $|x| > a$ corresponds to a disjunction.

Practice Exercises

1. 1, −2
2. C
3. −2, 2
   1, 3
4. 6, −2
5. C
6. B
7. C
8. D, C
9. A, F
10. −6, 10

Apply

How Do You Plan for a Healthy Aquarium?

This problem allows students to show that they can write both simple and compound inequalities to model real-world situations. Students also must apply these inequalities in their design of an aquarium.

- If time is limited, you may want to remove the requirements that students include pictures of the fish and descriptions of accessories for the aquarium.
- To ensure that students will need to write compound inequalities for the appropriate pH levels and water temperatures for their aquarium, you may want to require that they choose at least two different types of fish and that not all of their fish have the same ideal pH level or the same ideal water temperature.

The Apply problems are rich performance tasks that may have a variety of open-ended solutions. One sample student response for this problem can be found in the Model Lesson.

3.1 Understand and Interpret Functions

Check for Understanding: Investigation 1

1. C
2. B
3. A, B, C
4. III
5. For an algebraic model, each input value can have only one output value. Graph I has two output values for $x = -1$, and Graph II has multiple output values for $x = 1$.

Check for Understanding: Investigation 2

1. Sample response: For table A, the response in the x column must be any number not equal to −2, 1, or 0 and in the y column can be any number. For table B, the x column must contain −3, −2 or 1, and the y column must contain any number that doesn't match the one previously assigned to that x value.
2. A, C
3. B, C
4. The graph of a horizontal line does represent a function because no x value has more than one corresponding y value. The graph of the vertical line does not represent a function because every y value (output) has the same x value (input). Since x is always the same, there are multiple (an infinite number of) values for y.

Check for Understanding: Investigation 3

1. A. III, IV, V.
   B. I, II, VI.
2. A. I, III
   B. II, III
   C. I
   D. II
3. $\left\{\begin{matrix} D\text{: } -\infty < x < \infty \\ R\text{: } -1 \le y \le 1 \end{matrix}\right\}, \left\{\begin{matrix} D\text{: } -\infty < x < \infty \\ R\text{: } -\infty < y < \infty \end{matrix}\right\}, \left\{\begin{matrix} D\text{: } -2 < x < 2 \\ R\text{: } -1 < y \le 1 \end{matrix}\right\}$
4. The domain is all real numbers, or $\{D\text{: } -\infty < x < \infty\}$. The range is $y = 0$, because this is the only value of y for any given x. This line is the x-axis.

Understand and Interpret Functions *(continued)*

Check for Understanding: Investigation 4

1. A. –1
B. –1
C. 0

2. A. >
B. <
C. =
D. <
E. $x = -3$ or $x = 0$

3. A. $D(t) = 15t$
B. $D\left(\frac{1}{3}\right) = 15\left(\frac{1}{3}\right), D\left(\frac{1}{3}\right) = 5$
C. $12 = 15t, t = \frac{4}{5}$ hour or 48 minutes

Practice Exercises

1. A, B, D

2. A, C, E
B, D

3. A

4. 6

5. B

6. A

7. G, F

8. 6

9. 4
2.5
49
0, 18
4, 49

10. –8, 4

Apply

How Much Does a Million Dollars Weigh?

In this problem, students research the weight of a dollar bill and determine if there is a function relationship between the denomination and the weight of a million dollars.

The Apply problems are rich performance tasks that may have a variety of open-ended solutions. One sample student response for this problem can be found in the Model Lesson.

3.2 Analyze Arithmetic Sequences and Linear Functions

Check for Understanding: Investigation 1

1. 11, 32, 46, 60
A. 4
B. 7
C. NEXT = NOW + 7

2. 6, 10, 14, 18

3.

Shape x	Perimeter y
1	6
2	10
3	14
4	18

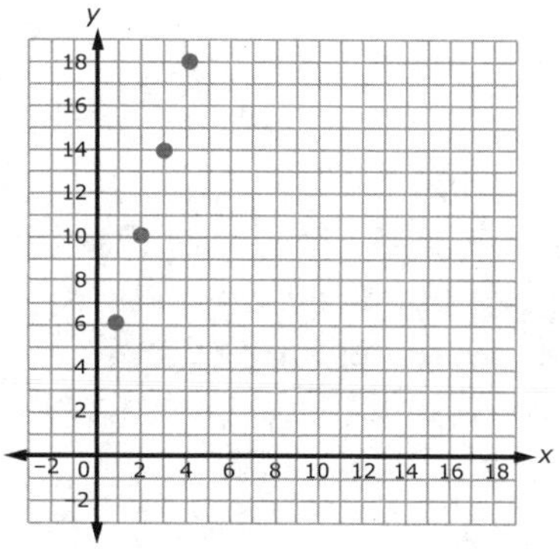

4. B, D, E

5. The start value is −2 for both sets, and the common difference for set A is 0, so this is an arithmetic sequence. But the difference for set B varies between 4 and −4, so this is not an arithmetic sequence.

Check for Understanding: Investigation 2

1. NEXT=NOW − 2

2.

Row (n)	Number of Cans $C(n)$
1	12
2	10
3	8
4	6
5	4
6	2

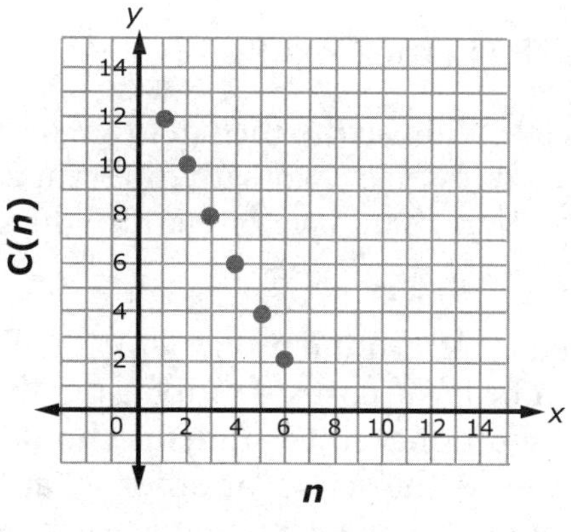

A. $C(1) = 12$; $C(n + 1) = C(n) - 2$
B. $C(n) = -2n + 14$ (or $C(n) = 14 - 2n$)

3. B, C

4. The y-intercept of the graph will move up by 2 to (0, 16) and the number of rows will increase by 1, so the new x-intercept will be (8, 0). The new function will be $C(n) = -2n + 16$ (or $C(n) = 16 - 2n$ or $C(n) = 16 + -2n$).

Analyze Arithmetic Sequences and Linear Functions *(continued)*

Check for Understanding: Investigation 3

1. $y = -\frac{2}{3}x - 2$
2. A. IV.
   B. VI.
   C. IV.
   D. V.
   E. IV.
   F. VI.
3. C, F
4. A. 120
   B. 50
   C. 70
5. The elevator travels at a rate of –12 ft/sec, representing the slope, which is negative because the elevator is going down. The graph's y-intercept would be (0, 840), representing the top floor height, and the x-intercept would be (70, 0), representing the time it reaches the ground at a height of 0. So the domain is real numbers from 0 to 70, and the range is real numbers from 0 to 840.

Practice Exercises

1. C
2. D
3. D
4. B
5. C, F
6. A
7. B
8. C
9. Andrew, Beatrice
10. A, B, D

Apply

How Many Seats Will Be in the Next Row?

In this problem, students will use the Internet to research seating plans at various stadiums and concert halls. They are instructed to select two with different arrangements and determine the number of seats in the subsequent row. They then have to identify the sequence and represent the data with implicit and explicit equations.

Students also are asked to determine profit, which will involve researching revenue gained from the additional seats and some basic costs of purchasing the new seats. To determine the average revenue that is generated from the additional seats, students should consider using the pricing information provided by searching the website for the venue, or other reliable online ticket agents. To find a basic cost for the actual seats being added to the venue, suggest that students use search phrases such as *stadium seat manufacturing* or *arena seat production*. Students also should research the average amount of use of each venue to approximate annual revenue.

Consider having students work in pairs to complete and present this problem.

The Apply problems are rich performance tasks that may have a variety of open-ended solutions. One sample student response for this problem can be found in the Model Lesson.

3.3 Analyze Geometric Sequences and Exponential Functions

Check for Understanding: Investigation 1

1.

Stage (n)	Number of Shaded Triangles $T(n)$
0	1
1	3
2	9
3	27

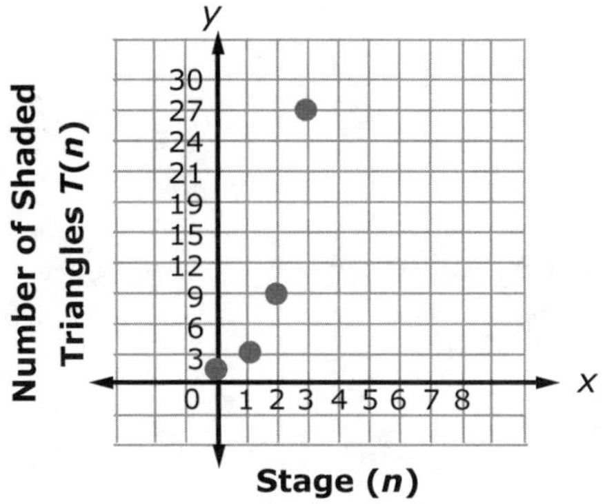

2. 3, 81

3. B, C, F

4. geometric, exponential

5. Both are correct. It is arithmetic because the common difference is 0, and it is geometric because the common ratio is 1.

Check for Understanding: Investigation 2

1.

Stage (n)	Approximate Total Area $A(n)$
0	8
1	6
2	4.5
3	3.38
4	2.53
5	1.9

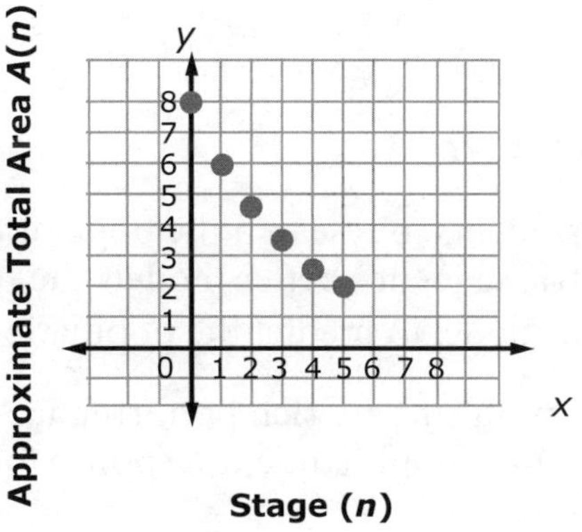

2. 0.75

3. A. NEXT = NOW · 0.75
B. $A(n + 1) = A(n) \cdot 0.75$

4. Stage 8

5. Since the ratio is the same for each successive term, this value is raised to increasing powers, beginning with 0. The start value multiplied by the common ratio raised to the term number will give the value of the sequence for that term number.

Analyze Geometric Sequences and Exponential Functions *(continued)*

Check for Understanding: Investigation 3

1. A. NEXT = NOW · 3
B. NEXT = NOW + 8

2. A. 18
B. 20

3. A. 2
B. 4
C. They would be the y-intercepts.

4. A. $L(n+1) = L(n) \cdot 3$
B. $L(n) = 2 \cdot 3^n$
C. $S(n+1) = S(n) + 8$
D. $S(n) = 8n + 4$

5. The ratios between the first three heights are all equal to 0.75. The 28 feet height is incorrect, and should be $36 \cdot 0.75 = 27$ feet. This will also make the last value of 20.25 feet correct.

Practice Exercises

1. A

2. 2, 0, 4.5

3. (0, 84)

4. A, C

5. E, C, A, F

6. A

7. A, D

8. B, C, D

9. B, C

Apply

How Does Your City Grow?

This problem allows students to research the population of a city and explore whether an arithmetic sequence or a geometric sequence better models the population growth. They will use either the geometric or algebraic model to predict the population the year they graduate from high school.

The Apply problems are rich performance tasks that may have a variety of open-ended solutions. One sample student response for this problem can be found in the Model Lesson.

4.1 Analyze Graphs of Functions

Check for Understanding: Investigation 1

1. A. 1.36 g / cm³, $m(V) = 1.36V$
B. 1.22 g / cm³, $m(V) = 1.22V$

2. A, C, D, F

3. The conversion factors from pounds to kilograms and feet to centimeters are not the same, so the value in customary units is different; however, when the units are converted, the values will be equivalent.

Check for Understanding: Investigation 2

1. A. $S(x) = 21x + 30$
B. $B(x) = 24x + 18$
C. y-intercept, slope

2. A. $S(2) = 72$
B. $B(2) = 66$

3. A. 8 days
B. 7 days

4. A. $2 \cdot S(x) + B(x) = 66x + 78$
B. $408

5. The y-intercept, or initial fee, for renting a snowboard is $18, which is less than the $30 for skis. The slope is greater, however, so the graph of the line will be steeper. The two lines intersect at $x = 4$ days, which is when the total cost is the same for both.

Check for Understanding: Investigation 3

1.

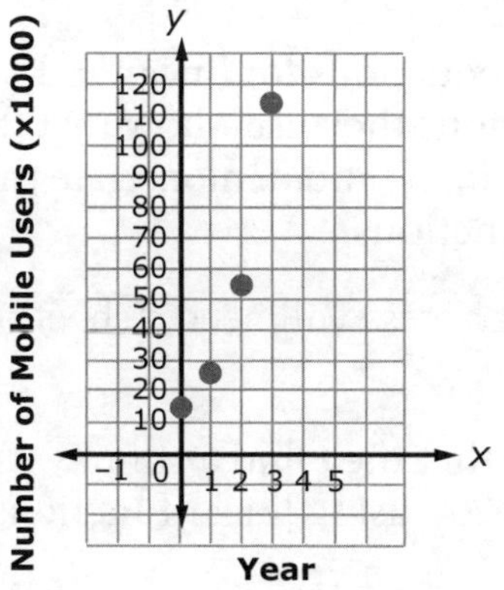

2. A. 2
B. 14
C. $M(x) = 14 \cdot 2^x$

3. 224

4. B, C, E

5. The initial value would change to 7 so the function would change to $M(x) = 7 \cdot 2^x$. The graph would shift right by 1 and down by 7.

Analyze Graphs of Functions *(continued)*

Check for Understanding: Investigation 4

1. A. III.
B. II.
C. I.
D. IV.

2. A. $y = \frac{2}{3}x - 2$
B. $y = -\frac{3}{2}x - 2$

3. perpendicular

4. A, B, D

5. The slope of line *s* is $\frac{b}{a}$, and any line perpendicular to it has slope $-\frac{a}{b}$, so beginning at *P*, count *b* units to the right and *a* units down to locate a point on line *t*. (Alternatively, count *b* units to the left and *a* units up.)

Practice Exercises

1. E

2. B, A, D, E, I

3. B

4. slope, 3, 9

5. $\frac{2}{3}$

6. B, C, F

7. C

8. B

9. −3

10. C

11. D

Apply

How Can You Use Functions to Make an Image?

This problem allows students to demonstrate that they can graph and write equations for linear and exponential functions. Students also must identify the domains of the functions they use, show that they understand the relationship between the slopes of parallel or perpendicular lines, and demonstrate the ability to identify and describe transformations of linear and exponential functions.

- Students who will be working on this problem at home without Internet access will need a sheet of graph paper or a printout of a coordinate plane for this problem.
- For students who have limited drawing ability, you may want to suggest that they find a simple line drawing online and then attempt to recreate it on a coordinate plane by using linear functions, exponential functions, and vertical segments.
- You may want to add a requirement that students use at least one linear function and at least one exponential function in their drawings.

The Apply problems are rich performance tasks that may have a variety of open-ended solutions. One sample student response for this problem can be found in the Model Lesson.

4.2 Compare Graphs of Linear and Exponential Functions

Check for Understanding: Investigation 1

1.

Table 1	
Round	**Number of Matches Left after Round**
Start of Tournament	32
1	16
2	8
3	4

Table 2	
Match	**Number of Players Left after Each Match**
Start of Tournament	128
1	126
2	124
3	122

2. A. geometric, ratio, 0.5 or $\frac{1}{2}$
B. arithmetic, difference, −2

3. A. $M(x) = 32 \cdot \left(\frac{1}{2}\right)^x$
B. $P(x) = -2x + 128$ or $P(x) = 128 - 2x$

4. A. {0, 1, 2, 3, 4, 5}, {1, 2, 4, 8, 16, 32}
B. integers 0 to 64, even numbers 0 to 128

5. A decreasing exponential function has a common ratio between 0 and 1 because each term is multiplied by a fraction, so its sign is positive. A decreasing linear function has a common difference that is negative because the same value is being subtracted each time. (Alternatively, the inverse of multiplication is division with the same sign and the inverse of addition is subtraction, or addition of the opposite.)

Check for Understanding: Investigation 2

1. A. linear, difference
B. exponential, ratio

2. A. $T(n) = 3n$
B. $S(n) = 3^n$

3. A. 18
B. 243
C. (0, 0), (0, 1)
D. 2, 9
E. 14

4. The function for Erik's quilt, $S(n)$, will have domain {1, 2, 3, 4, 5}, because this is the number of stages it will take to complete 243 squares, so the range is {3, 9, 27, 81, 243}. The function for Elena's quilt, $T(n)$, will have a domain of integers from 1 to 14, because this is the number of stages it will take to complete 42 triangles, so the range is multiples of 3 from 3 to 42.

Compare Graphs of Linear and Exponential Functions *(continued)*

Check for Understanding: Investigation 3

1. B, C, D

2. A. exponential, ratio, 2
B. linear, difference, 8

3. A. (0, 3)
B. (0, 3)

4. B, D

5. The domain of a sequence's function model is the term numbers, so the starting value of a sequence is the function value at $n = 1$. The function's initial value, or y-intercept, is its value at $n = 0$.

Practice Exercises

1. A

2. 0. 8, ratio

3. C

4. D

5. D

6. D

7. C

8. A. −1.5, 72
B. (48, 0)

9. 10, 32, B

10. $-2, \frac{1}{4}$

Apply

How Does a Fractal Grow?

This problem allows students to demonstrate their understanding of exponential functions as they identify an exponential relationship and represent it in different forms (diagram, table, graph, and equation). It also provides students with an opportunity to see how exponential functions can be useful for making predictions.

- Some common and relatively simple fractals that students might research include the Cantor set, the Koch snowflake, the H-tree or H-fractal, the Sierpinkski triangle, and the box fractal.
- Suggest that students draw Stage 0 of their fractal fairly large so that it will be easier to draw the later stages. Students can make their drawings by hand or by using the Dynamic Geometry Tool if they are familiar with its features.
- You may want to add the requirement that students explain what each parameter in the equation of their function represents in relation to the fractal.

The Apply problems are rich performance tasks that may have a variety of open-ended solutions. One sample student response for this problem can be found in the Model Lesson.

5.1 Solve Systems of Equations and Inequalities

Check for Understanding: Investigation 1

1. A. $y = 0.05x + 28$, $y = 0.07x + 25$
   B. $0.05x + 28 = 0.07x + 25$
   C. 150 miles
2. A. Equation: $5x - 3 = 3x - 5$, Solution: (–1, –8)
   B. Equation: $x - 6 = 4x - 3$, Solution: (–1, –7)
3. A. (1, 6) (Answers will vary but should be close to the exact solution.)
   B. $\left(\frac{3}{4}\right), \left(\frac{23}{4}\right)$, or (0.75, 5.75)

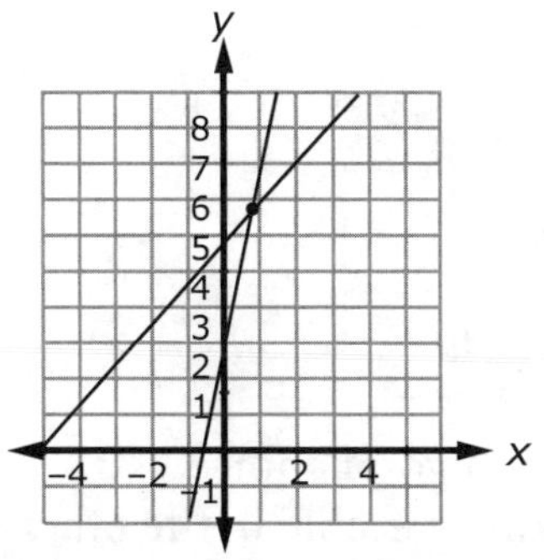

4. The solution to the equation is the value of x for which the y values are equal. Graphically, this is the x value of the point on the graph that is the intersection of the two lines.

Check for Understanding: Investigation 2

1. 1.25 hours, 2 hours, never
2. A. A, B
   B. 8 miles
   C. 4 mph
3. A. II.
   B. III.
   C. I.
4. To create a system with no solutions, the lines must be parallel with different y-intercepts. So the second line should have the same value for m, but a different value for b. For infinitely many solutions, the second line must be a multiple of the first, so each of the terms must be multiplied by the same value.

Solve Systems of Equations and Inequalities *(continued)*

Check for Understanding: Investigation 3

1. A. $324x + 670y = 22{,}266$, $486x + 1{,}255y = 33{,}849$

   B. -1.5 or $-\frac{3}{2}$

   C. $250y = 450$

   D. 65, 1.8

   E. 64 GB

2. A. III.
   B. I.
   C. II.

3.

Systems	Multiply Equation by	Equivalent systems
$3x + 4y = 8$	• 2 (or 3)	$6x + 8y = 16$ (or $15x + 12y = 24$)
$2x + 3y = 8$	• −3 (or −4)	$-6x - 9y = -12$ (or $-8x - 12y = -16$)

4. If one of the variables has the number 1 as a coefficient, then substitution is a good choice. If a coefficient for one of the variables is a multiple of the coefficient for the same variable in the other equation, then only one multiplication is necessary. If neither coefficient is a rational multiple of the corresponding coefficient in the other equation, then two multiplications will be necessary.

Check for Understanding: Investigation 4

1. A. $6x + 9.5y > 300$
   B. $x + y \leq 80$

2. A. $y > -\frac{1}{2}x + 2$
   B. $y \geq 2x - 5$

3. A. $y \geq -1$
   B.

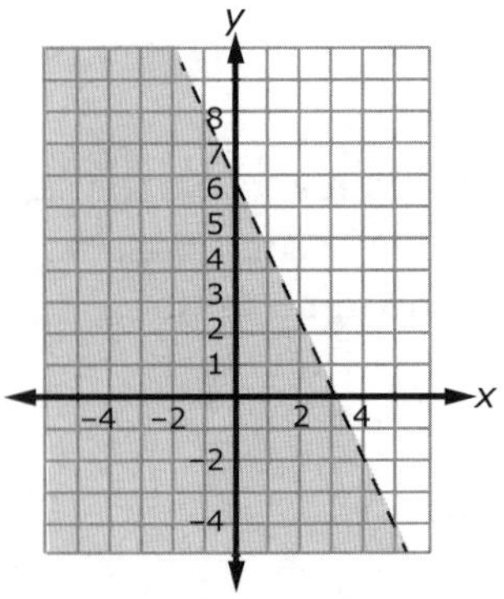

4. When the inequality is strict (< or >), the line is dashed because it does not include the points on the line. The line is solid when the inequality is inclusive (≤ or ≥). For any given value of x, all y values that solve the inequality are either greater than the expression and above the y value on the line so the graph should be shaded above, or less than the expression and below the y value on the line so it should shaded below the line (Alternatively, points on, above, and below the line can be checked for whether they solve the inequality.)

Solve Systems of Equations and Inequalities *(continued)*

Check for Understanding: Investigation 5

1. A. $10x + 15y \geq 180, x + y \leq 15$
B. (2, 12), (5, 9), (7, 8), (0, 12)

2. A. $\begin{cases} 3x + 2y \leq 20 \\ x + y < 8 \end{cases}$ or $\begin{cases} y \leq -\frac{3}{2}x + 10 \\ y < -x + 8 \end{cases}$

B.

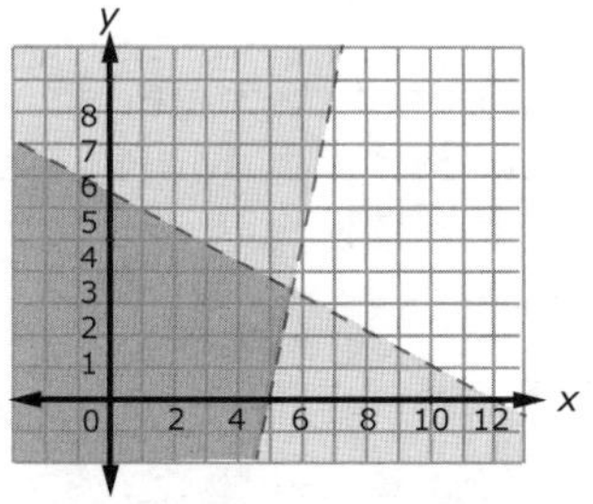

3. The x-intercept represents the maximum x value that fits both constraints, which means that the y value is 0. The y-intercept represents the maximum y value that fits both constraints, which means that the x value is 0. The minimum domain and range values in a context such as a craft fair are $x = 0$ and $y = 0$ because you can't make a negative number of items.

Practice Exercises

1. $\frac{3}{4}, \frac{5}{6}$

2. 3, –5

3. A. 18
B. 24
C. (4, –1)

4. C, B, F

5.

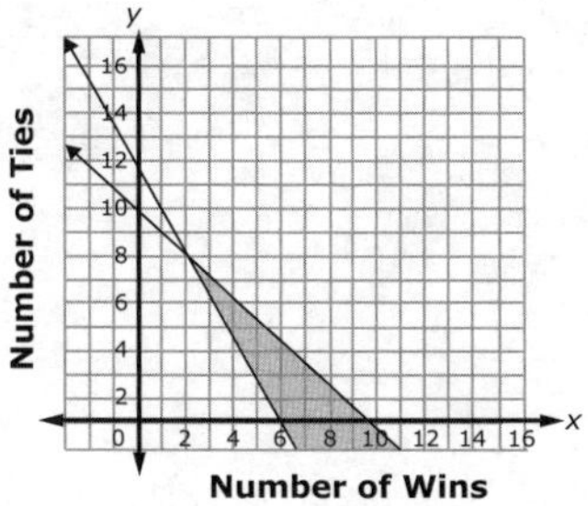

6. (6, 8)

7.

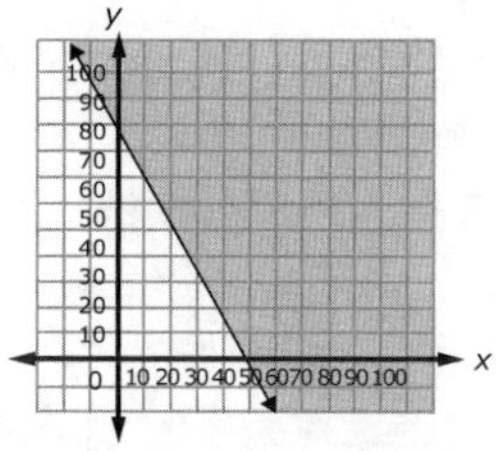

8. E

9. A, C

10. C

11. D

Solve Systems of Equations and Inequalities *(continued)*

Apply

Can Selling the Pizza Bring in the Dough?

This problem provides students opportunities to apply what they learned to write and graph systems of equations to represent a real-world situation, and interpret the graphs in the context of the problem.

- If research is not an option, students may estimate the costs, or use these figures: $10.95 for frozen pizza, $16 for restaurant pizza, $6.77 for 200 paper plates, and $3.01 for napkins.
- Students can graph the equations by hand or use the Graphing Calculator.

The Apply problems are rich performance tasks that may have a variety of open-ended solutions. One sample student response for this problem can be found in the Model Lesson.

5.2 Use Systems in Decision Making: Linear Programming

Check for Understanding: Investigation 1

1. A. $t \geq c$ (or $c \leq t$)
   B. $t \leq 14$
   C. $c + t \geq 18$
   D. $c + t \leq 22$
   E. $C = 16c + 12t$
2. A.

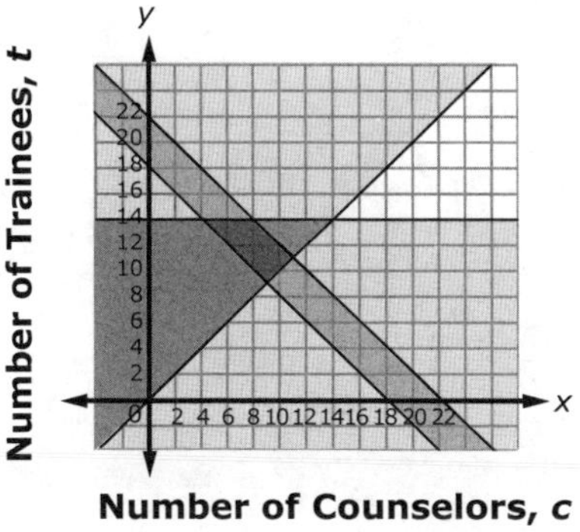

   B. I. (4, 14), II. (8, 14), III. (9, 9), IV. (11, 11)
3. A. 4
   B. 14
4. The integer coordinates that are nearest to the vertex and within the feasible region should be used.

Check for Understanding: Investigation 2

1. A. ounces, ounces
   B. $1 \leq x \leq 2$, $y \geq 1$, $30x + 120y \leq 200$, $4x + 6y \geq 12$
   C. $37.5x + 26.8y$
2.

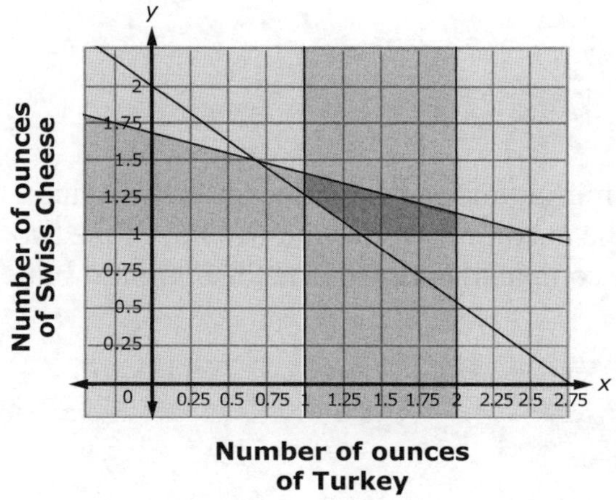

3. 1, 1, 1.5, 2, 2
4. 73.14¢ (or 73¢ or \$0.73)
5. The values of the vertices solve the whole system and they represent the minimum or maximum for each pair of lines, so the values that minimize or maximize the objective function must be among these vertices.

Use Systems in Decision Making: Linear Programming *(continued)*

Check for Understanding: Investigation 3

1. number of this year's adult dogs, number of this year's puppies
2. $x \geq 0, y \geq 0$
   $x \leq 16, y \leq 12$
   $600x + 800y \leq 12{,}000$, $1.2x + 0.8y \leq 16.8$
3. A.

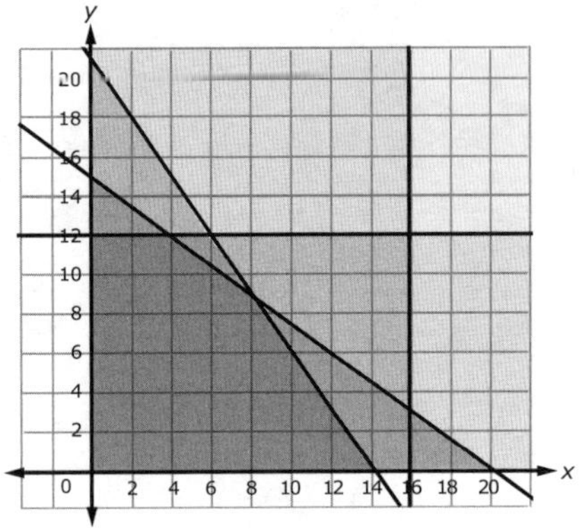

B. (0, 0), (0, 12), (4, 12), (8, 9), (14, 0)
C. $3500x + 2500y$
D. 8, 9, $50,500

4. Yes, this is possible, because the constraints could make a feasible region with vertices that lie only on the axes. In this case, the value for each of the parameters in the scenario will be either 0 or its maximum value.

Check for Understanding: Investigation 4

1. A. VIII.
   B. I.
   C. XI.
   D. IV.
   E. X.
2. 12
3. The points that lie on each line represent the maximum or minimum function value for each x value. Therefore, the intersection of each pair of lines represents the maximum or minimum value of the two functions creating that vertex, so these are the ones that are compared to determine the optimum value for the entire system.

Use Systems in Decision Making: Linear Programming *(continued)*

Practice Exercises

1. A, B, D
2. A, E
3. $11.80
4. 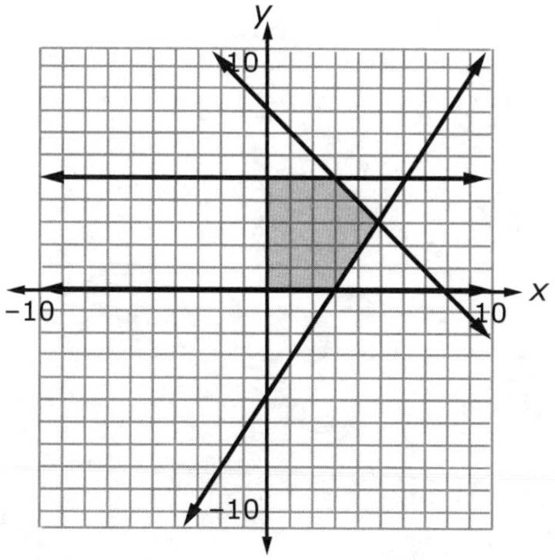

5. $2.20, 2, 1
6. 760, 30, 20
7. $x + y, \leq 2y, \geq 8, \geq 0, \leq 15$
8. 210, 0, 6
9. $2,100, 60, 40

Apply

How Much Solar Power Should Your Community Purchase?

This problem allows students to demonstrate their ability to use linear programming to minimize a value in a real-world problem. It also challenges students to use mathematical reasoning to identify reasonable constraints.

- Allow students to have access to the Graphing Calculator for this problem. They will need to graph the constraints in this linear programming problem in order to determine the feasible region.
- As an extension, students could solve this problem by researching and using actual data about the electric utility or one of the electric utilities in their own community. Students would need to find or calculate the utility's average production cost per megawatt-hour of electricity, the number of megawatt-hours of energy its customers use per year, and the average CO_2 equivalent generated by the utility's power sources per megawatt-hour. Depending on the electric utility, this information may or may not be readily available. If it is not easily accessible, suggest an alternate community or utility for students to use. One possibility is New York State Electric and Gas (NYSEG).

The Apply problems are rich performance tasks that may have a variety of open-ended solutions. One sample student response for this problem can be found in the Model Lesson.

6.1 Represent and Analyze Data

Check for Understanding: Investigation 1

1. symmetric, skewed left
2. A, C, D, G
3. An extreme data point will move the mean in the direction of the point's value, so the median is a better measure of center. Removing it will move the mean closer to the median so that either will reflect the central tendency. Because variability is dependent on the distribution for all the data, removing the extreme value will have little effect on IQR, but it will change the spread or range.

Check for Understanding: Investigation 2

1. A. 18, 18
   B. 22, 21
2. A. less than
   B. less than
   C. 50%, 26%
   D. 77.8%, 32%
   E. MAD, IQR
   F. hobby
3. Even though the jewelry store employees work an average of 3 hours less per week than the hobby store employees, their hours are more consistent, so Melanie should work for the jewelry store to meet her goal.

Check for Understanding: Investigation 3

1. A. I.
   B. VIII.
   C. I.
   D. III.
   E. II.
   F. VII.
   G. IV.
   H. VI.
2. A. 62.5%
   B. 62.5%
   C. symmetric
   D. skewed left
   E. popcorn prices
3. If the mean and median are close together, the standard deviation will be smaller in value because more of the points are near the center of the set, and it can be used to describe the variation of the set. If the mean and median are far apart, the standard deviation will be greater because of the skew and is therefore not a good measure of variability.

Represent and Analyze Data *(continued)*

Check for Understanding: Investigation 4

1.

	Mean Distance (m)	Median Distance (m)	Standard Deviation	IQR
Spudnik	245.6	240.5	23.1	24
Later Tater	243.5	243.5	12.9	22.5

2. B, D, E

3. A. skewed right, left of
B. symmetric, directly in

4. Since the data for Later Tater is more consistently near its median, and the data for Spudnik is more variable, the class should choose Later Tater. It will have the greater chance of exceeding the minimum in three of five launches.

Practice Exercises

1. B

2. C, B, A

3. A, A, B

4. B, D

5. A, C, D

6. E

7. B, C, E

8. A, B

9. C

Apply

How Should Data Direct Tornado Research?

In this problem, students determine the most appropriate type of display to present data and justify their choices. Then, they use measures of center and variability to compare the data and interpret the results.

- Students who do not have Internet access at home may need to watch the Studying Tornadoes video and complete the research at school.
- If students work in pairs, each can analyze one state, and they can check each other's work. Then, they can compare the data for the two states together.
- For background, you may wish to discuss which states are most strongly affected by tornadoes. Although they can occur in any state, they occur most frequently in an area of the central United States commonly known as Tornado Alley.
- As an extension, combine the data for the entire class and compare all of the states that were chosen.

The Apply problems are rich performance tasks that may have a variety of open-ended solutions. One sample student response for this problem can be found in the Model Lesson.

6.2 Analyze Scatter Plots

Check for Understanding: Investigation 1

1. B, D, E, F

2. A. III.
B. I.
C. IV.
D. II.

3. A. the slope
B. the slope
C. the correlation coefficient
D. the same
E. both values

4. Since the graph of the line is decreasing, every hour of exercise is associated with a decrease of about 4 beats per minute in the resting heart rate. The value of the correlation coefficient indicates that there is a strong negative linear correlation between hours of exercise per week and heart rate.

Check for Understanding: Investigation 2

1. A. difference, y value, y value
B. vertical
C. near zero

2.

x	2	5	7	8	11	12
y	1.5	9.8	18.9	20.2	30.1	32.7
Residual	0.08	−1.13	1.63	−0.24	0.15	−0.42

3. C

4. A decreasing linear function will have an x-intercept, meaning that according to the model, winning times will continue to decrease until they reach 0, which is impossible. (There will be a limit to the fastest time that will necessarily be greater than 0.)

Check for Understanding: Investigation 3

1. B, C, E

2. C, E

3. There is a strong correlation between distances and airfares for the data they collected. However, they have only a few examples that could be from limited carriers and that do not account for other variables, such as airport fees and markets with greater or lesser demand.

Analyze Scatter Plots *(continued)*

Check for Understanding: Investigation 4

1. B, D, E

2. A. Playing sports is related to higher grades among students.
B. Maintaining strong social ties is connected to better long-term health.
C. A decreased risk for some cancers linked to eating Italian pizza each week.

3. tied, tends, linked, related, associated

4. A strong correlation is a linear measure of association only and can never prove causation. To prove causation, scientific studies that control for other variables must be conducted.

Practice Exercises

1. 4

2. 87%

3. D, E

4. A, E, F

5. A, E

6. A

7. Yes, No

8. 147

9. D

10. C

11. B, F

12. C

Apply

How Do Populations Change over Time?

In this problem, students use data to model population growth with a linear function. They then use residuals to determine whether the model is appropriate, and predict the population in 10 years.

- Students who do not have Internet access at home may need to watch the video at school.
- For each continent, you may wish to select a student who chose that continent to present his or her findings to the class. Then, lead a class discussion comparing the population trends on all of the continents.

The Apply problems are rich performance tasks that may have a variety of open-ended solutions. One sample student response for this problem can be found in the Model Lesson.

6.3 Interpret Two-Way Frequency Tables

Check for Understanding: Investigation 1

1.

	Drive Themselves	Use Other Transportation	Total
Juniors	48	80	128
Seniors	90	22	112
Total	138	102	240

2.

	Drive Themselves	Use Other Transportation	Total
Juniors	0.20	0.33	0.53
Seniors	0.38	0.09	0.47
Total	0.58	0.43	1.00

3. A. IV. B. III.
   C. I. D. II.

4. The ratio of two values in one table will be equivalent to the same ratio in the other table. Among all juniors, for example, the conditional relative frequency of those who drive themselves to school is $\frac{48}{128} = 0.375$, using the frequency table, or $\frac{0.2}{0.533} \approx 0.375$, using the relative frequency table, with minor differences possible due to rounding.

Check for Understanding: Investigation 2

1.

		Fall Festival		
		Yes	No	Total
Spring Fling	Yes	0.25	0.33	0.58
	No	0.29	0.13	0.42
	Total	0.54	0.46	1.00

2. A. IV. B. VI.
   C. II. D. V.
   E. I. F. III.

3. Overall, 58% voted yes to Spring Fling, and 57% of those also voted no to Fall Festival. By comparison, 54% overall voted yes to Fall Festival, and 54% of that group voted no to Spring Festival. Since more voted *for* Spring Fling *and* fewer voted *against* Fall Festival, Spring Fling would be the better choice.

Check for Understanding: Investigation 3

1.

	11 p.m.	12 a.m.	1 a.m.	Later or No Curfew	Total
9th Grade	26	52	46	4	128
10th Grade	18	62	42	14	136
11th Grade	6	28	30	32	96
12th Grade	0	8	28	52	88
Total	50	150	146	102	448

2. A. conditional relative B. marginal relative
   C. joint relative D. 21%
   E. 77% F. 11th

3. Already, 65% of 11th graders can stay out until at least 1 a.m. Of those 62 students, 32 can stay out even later (greater than half!), and many of those quite possibly past 2. I am asking only for the hour extension to 2 a.m.

Interpret Two-Way Frequency Tables *(continued)*

Practice Exercises

1. B, D, F
2. 0.31, 0.15, 0.25, 0.11
3. 9–11 hours, B, 179
4.

	D	B	Total
A	15	20	35
C	31	10	41
Total	46	30	

5. 0.27, 0.22, 0.31
6. Marginal: A, F
   Joint: B, D
   Conditional relative frequency: C, E
7. D
8. 0.45, 0.45

Apply

Don't You Just Love the Winter Olympics?

This problem provides students the opportunity to write and answer their own questions that can be answered using a two-way frequency table. The questions must include joint, marginal, and conditional relative frequencies.

- Students who do not have Internet access at home may need to watch the video and complete the research at school.
- As a challenge, students can pair up and try answering each other's questions.

On the Winter Olympics Data sheet, the values have been rounded. Because of this, it is possible that the sum of the values may not exactly match the totals on the sheet.

The Apply problems are rich performance tasks that may have a variety of open-ended solutions. One sample student response for this problem can be found in the Model Lesson.

7.1 Explore Normal Distributions

Check for Understanding: Investigation 1

1. 68, 95, 99.7
2. B, C
3. The mean of the data is 12.36 cm; the median is 12 cm; the mode is 12 cm; and the standard deviation is about 1.45 cm. The data does appear to approximate a normal distribution, because the mean, median, and mode are all about the same. Also, $\frac{34}{50} = 0.68$ = 68% of the wingspans are within 12.36 ± 1.45 cm, $\frac{48}{50} = 0.96$ = 96% are within $12.36 \pm (2)(1.45)$ cm, and $\frac{50}{50} = 1.00$ = 100% are within $12.36 \pm (3)(1.45)$ cm, as would be expected.

Check for Understanding: Investigation 2

1. A, E
2. right, left
3. −0.75, 22.7
4. B
5. A pumpkin weighing 20 pounds has a *z*-score of about $z = \frac{20 - 18}{3} = 0.67$, while one weighing 13 pounds has a *z*-score of about $z = \frac{13 - 18}{3} = -1.67$. According to a *z*-table, the area under a normal curve to the left of $z = 0.67$ is 0.7486, and the area to the left of $z = -1.67$ is 0.0475, so the area between $z = 0.67$ and $z = -1.67$ is $0.7486 - 0.0475 = 0.7011$. Therefore, approximately 70.11% of the pumpkins weigh between 13 and 20 pounds.

Explore Normal Distributions *(continued)*

Practice Exercises

1. A

2. 34

3. C

4. C

5. Lisa's z-score: 1.5
Matt's z-score: 0

6. $L = 4 \cdot 90 - (x + y + z)$

7. 1.6

8. 9

9. 10

10. 84

11. C

Apply

Which Score is Better?

This problem provides students opportunities to analyze the scores on the ACT and SAT and draw conclusions about the likelihood of certain scores being earned.

- If research is not an option, you may wish to provide this data from 2010:
    - For the ACT, the mean composite score was 21.0 with a standard deviation of 5.2.
    - For the SAT critical writing score, the mean was 501 with a standard deviation of 112.
    - For the SAT mathematics score, the mean was 516 with a standard deviation of 116.

The Apply problems are rich performance tasks that may have a variety of open-ended solutions. One sample student response for this problem can be found in the Model Lesson.

7.2 Collect, Analyze, and Interpret Statistical Data

Check for Understanding: Investigation 1

1. $\frac{1}{70}$
2. heterogeneous, homogeneous
3. D
4. It depends on how the list of the association's members is organized. For example, if the list is organized in alphabetical order, then Jamal used systematic sampling correctly, and the mean salary he finds is likely to be representative of the population. However, if the list is organized by academic department, then this is probably not the case, because professors in some departments might earn a lot more than those in others.

Check for Understanding: Investigation 2

1. A, E, F
2. 3
3. The fraction of 6-8 splits picked up is $\frac{29}{174} = \frac{1}{6}$, the fraction of 6-8-10 splits picked up is $\frac{38}{342} = \frac{1}{9}$, and the fraction of 8-10 splits picked up is $\frac{7}{252} = \frac{1}{36}$. Two number cubes can come up in 36 possible ways. Of these, 6 result in a sum of 7; 4 result in a sum of 5; 4 result in a sum of 9; 1 results in a sum of 2; and 1 results in a sum of 12. Therefore, a sum of 7 could appropriately represent picking up a 6-8 split, a sum of 5 or 9 could represent picking up a 6-8-10 split, and a sum of 2 or 12 could represent picking up an 8-10 split.

Check for Understanding: Investigation 3

1. D
2. A, D, F
3. 0.0153
4. wider
5. The proportion of listeners in the sample who preferred jazz to classical is $\hat{p} = \frac{165}{250} = 0.66$. The population proportion is unknown, so it's necessary to use the standard error, which is $\sqrt{\frac{0.66(1-0.66)}{250}} \approx 0.0300$. Since a 95% confidence interval is ±2 standard errors from the sample proportion, at a 95% confidence level, the percentage of all of the station's listeners who prefer jazz to classical is $0.66 \pm 2(0.0300) = 0.66 \pm 0.0600 = 66\% \pm 6\%$.

Check for Understanding: Investigation 4

1. C
2. A, C
3. The mean length of the vacations of the 50 tourists in the sample is 9.1 days, and the standard deviation is about 1.75 days. Using the standard deviation and the sample size, the standard error can be calculated as $\frac{1.75}{\sqrt{50}} \approx 0.25$ days. Since a 95% confidence interval is ±2 standard errors from the sample mean, at a 95% confidence level, the mean length of the vacations of all tourists who visit the island is $9.1 \pm 2(0.25) = 9.1 \pm 0.5$ days.

Collect, Analyze, and Interpret Statistical Data *(continued)*

Check for Understanding: Investigation 5

1. observational study, experiment
2. D
3. F, G, H
4. If the company's goal is to bring in as much revenue as possible from the shoes, then it should charge $69.99. This is because for every 100 users who are shown a price of $49.99, it will bring in $\$49.99 \cdot 17 = \$849.83$, while for every 100 users who are shown a price of $69.99, it will bring in $\$69.99 \cdot 13 = \$909.87$. However, maybe the company is overstocked and wants to sell as many shoes as possible. If so, $49.99 is the better price.

Practice Exercises

1. A
2. B
3. biased
4. B
5. 1.32
6. D
7. $\frac{282}{642}$
8. 2
9. B

Apply

How Many Letters Are in an Average Word?

This problem allows students to demonstrate their ability to use and compare different sampling techniques to make and verify a conjecture about the average length of words in a passage.

- Before the activity begins, the class must determine how to handle numbers in the text, hyphenated words, proper names, and any other anomalies that may affect the results.
- One thing that may come up in the comparison is the sample size each used. Students may also consider combining their results from one type of sampling to see how the combination compares to the individual results.

The Apply problems are rich performance tasks that may have a variety of open-ended solutions. One sample student response for this problem can be found in the Model Lesson.

8.1 Create and Analyze Piecewise Functions

Check for Understanding: Investigation 1

1. A. $0 \leq x \leq 120$
   B. $0 \leq y \leq 35{,}000$
2. A. IV.
   B. V.
   C. I.
   D. VI.
   E. II.
   F. III.
3. A. 35,000
   B. 30,000
   C. 25,000
4. A point will actually fit in each function definition on either side when it joins the segments, but it is in the domain of the function piece that is closed (defined by ≤ symbol) at the end of the interval on which it is defined at that point.

Check for Understanding: Investigation 2

1.

Rescue Team's Elevation, x	2,925	3,290 or 4,460	3,570 or 4,180	4,430
Vertical Distance from Climber, y	950	585	305	555

2. A, B, D
3. C
4. left 5, down 3
5. The minimum point is (h, k) and as a piecewise function this can be written as:

$$f(x) = \begin{cases} -x + h + k, \text{ if } 0 \leq x < h \\ x - h + k, \text{ if } h \leq x \leq c \end{cases}.$$

Create and Analyze Piecewise Functions *(continued)*

Check for Understanding: Investigation 3

1. A. $0 < x < 4$
B. $4 \leq x < 6$
C. $6 < x \leq 7$

2.

$$f(x) = \begin{cases} 2x^2, \text{ if } 0 \leq x < 3 \\ -2x + 24, \text{ if } 3 < x \leq 4 \\ 2x, \text{ if } 4 < x \leq 8 \end{cases}$$

(or $3 < x < 4$ and $4 \leq x \leq 8$)

3. B, D

4. A discontinuity can be at a shared interval endpoint that is open, so it is not included in either interval, or it can be due to different values for x at the endpoints for adjacent intervals. The set of equations is a function if there are no overlaps in the domain intervals, including the sharing of closed interval endpoints.

Practice Exercises

1. $1|x - 4| + 0$

2. 14

3. $f\begin{cases} 1 & \text{if } 0 < x \leq 3 \\ 3 & \text{if } 3 < x \leq 6 \\ 5 & \text{if } 6 < x \leq 9 \end{cases}$

4. $-|x - 1| - 3$

5. D

6. A. $x + 1$
B. $-x + 19$

7. $T(x) = \begin{cases} 0.10x & \text{if } 0 < x < \$8,926 \\ 0.15x & \text{if } \$8,926 \leq x < \$36,251 \\ 0.25x & \text{if } \$36,251 \leq x \leq \$87,850 \end{cases}$

8. Provider A: 35, 85
Provider B: 45, 75

9. $-|x + 3| + 2$

10. C, D

Apply

How Many T-Shirts Should You Purchase for a Fundraiser?

This problem provides students opportunities to use piecewise functions to solve a real-world problem. Students use piecewise functions to model the cost of buying T-shirts and graph the functions to compare the costs of different companies.

Students could use the Graphing Calculator, Whiteboard, or graph paper and pencil to graph the piecewise functions.

The Apply problems are rich performance tasks that may have a variety of open-ended solutions. One sample student response for this problem can be found in the Model Lesson.

8.2 Investigate Square Root and Cube Root Functions

Check for Understanding: Investigation 1

1. (0, 0), (0, 0)
2. range, $[0, \infty)$ (or $x \geq 0$), (b, a)
3. range, (∞, ∞) (or all real numbers), (d, c)
4. Raising a negative number to an odd power results in a negative number, meaning that a negative number has an odd root, so these values can be included in the domain for the odd root function. A negative number raised to an even power is positive, so all values for the domain of an even root function must be non-negative.

Check for Understanding: Investigation 2

1. A. right, 6, down, 3
   B. left, 4, up, 2
   C. −7, −5
2. A. $[0, \infty)$ (or $x \geq 0$)
   B. $(-\infty, \infty)$ (or all real numbers)
   C. $[h, \infty)$ (or $x \geq h$), $[k, \infty)$ (or $y \geq k$)
   D. $(-\infty, \infty)$ (or all real numbers), $(-\infty, \infty)$ (or all real numbers)
3. A, D
4. On the interval $(1, \infty)$, the square root of a number will be greater than the cube root. On the interval (0, 1), the square root function will be less than the cube root because you are dividing by a larger number. For example, $\sqrt{8} = 2\sqrt{2}$ and $\sqrt[3]{8} = 2$, but $\sqrt{\frac{1}{8}} \approx 0.35$ and $\sqrt[3]{\frac{1}{8}} = \frac{1}{2}$.

Check for Understanding: Investigation 3

1. (0, −100), depth before ascent (or equivalent)
2. $f(x) = 9x - 100$
3. $f(x) = 20\sqrt{x} - 100$
4. 9, 4
5. 11 (or 11.1), 25
6. The graph would be of a piecewise function with intervals that alternate between segments having a slope of 9 for each portion of ascent, and segments having a slope of 0 for decompression stops.

Investigate Square Root and Cube Root Functions *(continued)*

Practice Exercises

1. D
2. B, D, E
3. C, E
4. C, D
5. A, B, E
6. B
7. E, H
8. A, H
9. (−4, −3)
   (0, −1)
   (5, 0)
   −4
   −3
10. B

Apply

How Far Can You See?

This problem allows students to discover the relationship between the height of a person's eyes from the ground and the distance they can see to the horizon. Students model the relationship with a square root function. Students graph the function and interpret the graph to solve the problem.

- If research is not an option, you may wish to suggest students estimate a familiar building's height. Or they might consider the Goldman Sachs Tower in Jersey City, New Jersey, which is 781 feet tall.
- If you use this problem as a homework assignment, you may want to set up the problem during class time by having students watch the video Apply Problem: The Horizon.

The Apply problems are rich performance tasks that may have a variety of open-ended solutions. One sample student response for this problem can be found in the Model Lesson.

8.3 Investigate Rational Exponents

Check for Understanding: Investigation 1

1. radicand, exponent, index
2. index, radicand, exponent
3. B, F
4. $\sqrt{6}, \sqrt[n]{10}, (\sqrt[5]{m})^3$
5. $3^{\frac{1}{4}}, 5^{\frac{3}{2}}, p^{\frac{q}{r}}$
6. Yes, they are equal. The first expression can be written as $a^{\frac{m}{n}}$, which is then equivalent to $(a^m)^{\frac{1}{n}}$. Rewriting this expression back in radical form, we get $\sqrt[n]{a^m}$.

Check for Understanding: Investigation 2

1. A. IV.
   B. III.
   C. V.
   D. I.
   E. II.
2. A, C, D
3. $2^{\frac{11}{4}}, 3^{\frac{5}{6}}$
4. First, I can rewrite 4 as a power of 2 and then apply the power of a power rule to write the second part of the expression: $4^{\frac{1}{3}} = (2^2)^{\frac{1}{3}} = 2^{\frac{2}{3}}$. Now I can apply the power of products rule and then simplify: $2^{\frac{4}{3}} . 2^{\frac{2}{3}} = 2^{\frac{6}{3}} = 2^2 = 4$. (Students might also rewrite the first expression as $2^{\frac{4}{3}} = (2^2)^{\frac{2}{3}} = 4^{\frac{2}{3}}$ and them multiply to get $4^{\frac{2}{3}} \cdot 4^{\frac{1}{3}} = 4$.)

Check for Understanding: Investigation 3

1. A. $\sqrt{2} \cdot \sqrt{5}$
   B. $\sqrt{3 \cdot 7}$ (or $\sqrt{21}$)
2. C
3. A. $\sqrt{36 \cdot 2} = 6\sqrt{2}$

   B. $\sqrt{16 \cdot 6} = 4\sqrt{6}$

   C. $\sqrt{4 \cdot 30} = 2\sqrt{30}$
4. Writing the radicand in its prime factors will show how the perfect squares can be grouped. Since $108 = 4 \cdot 27 = 2^2 \cdot 3^3$, you can see a perfect square of 4 and of 9 with a 3 left over: $\sqrt{108} = \sqrt{2^2 \cdot 3^2 \cdot 3} = 2 \cdot 3\sqrt{3} = 6\sqrt{3}$. This method might be useful for a larger number when it is harder to see it as the product of perfect squares.

Investigate Rational Exponents *(continued)*

Practice Exercises

1. B, D, A, C
2. $4\sqrt{2}$
3. A, C, G, H
4. 31
5. A, D
6. A
7. B
8. 25
9. B, E, F
10. 36
11. $\frac{9}{16}$
12. 4
13. 4
14. $4\sqrt{6}$

Apply

How Do Consumer Prices Change over Time?

This problem allows students to demonstrate their ability to work with exponential functions by using a formula to compare the rates of two quantities increasing exponentially.

- Encourage students to use different products, so they can compare and contrast the average rates of change of many product prices through a class discussion after the assignment.
- If research is not an option, you may wish to provide students with this data:

Product	January 1913 Average Price (in dollars)	January 2013 Average Price (in dollars)
milk, per gallon	0.356	3.526
eggs, per dozen	0.373	1.933

The Apply problems are rich performance tasks that may have a variety of open-ended solutions. One sample student response for this problem can be found in the Model Lesson.

9.1 Represent Exponential Functions

Check for Understanding: Investigation 1

1. C, F
2. A. II, IV
   B. I, III
   C. IV
   D. III
3. 2^x
4. The y-intercept of 500 is the number of bacteria when you start counting the function at time $t = 0$.

Check for Understanding: Investigation 2

1. C
2. B, C
3. $a^x + 88$
4. A. II.
   B. III.
   C. I.
5. Average rates of change are small at first but increase. The graph becomes steeper as x increases.

Check for Understanding: Investigation 3

1. 3, 21, 78
2. A. II.
   B. III.
   C. I.
3. $-\infty < y < \infty$, $y \leq 0$, $1 < y < \infty$
4. Because the rate of change switches sign, it cannot match either a linear or exponential model. They are either always negative rates of change or always positive rates of change. The data could fit a quadratic model, like $y = (x - 5)^2$.

Represent Exponential Functions *(continued)*

Practice Exercises

1. B

2. Vertical stretch: 5
Vertical translation: –7

3. B

4. B, D (either order)

5. $a = 2, b = 1.4, c = 1$

6. D

7. B, F

8. 8.65

9. Domain: all real numbers
Range: $y < -4$

10. C

Apply

How Much Will You Need to Save for Your Child's College Tuition?

In this problem, students use a monthly payment function to determine how much they would need to save each month for a child's college tuition. They will compare the amount needed to save for three different start times, for a public and a private college, and explore alternatives to a four-year college education.

- Remind students that the percentage rate must be expressed as a decimal, for example 3% as 0.03, and that the annual percentage rate must be divided by 12 to get the monthly percentage rate.
- For background, you may wish to explain the parts of the payment formula and how they relate to each other. Compare this function to the compound interest formula: if P is the principal and r is the annual percentage rate divided by 12, expressed as a decimal, then the amount saved in t months is $A(t) = P(1 + r)^t$. This represents a single payment that accrues interest over time. For the monthly payment function, the same amount is added to the account each month, so each payment accrues interest for one month less than the previous payment.
- As an extension, students can explore the following financial aid alternatives: research functions for calculating what payments are on student loans and compare the total amount spent for saving in advance to taking out a loan.

The Apply problems are rich performance tasks that may have a variety of open-ended solutions. One sample student response for this problem can be found in the Model Lesson.

9.2 Analyze Exponential Growth and Decay Models

Check for Understanding: Investigation 1

1. B
2. B
3. \$3,163, \$3,546
4. The variable *a* represents the original amount invested. The variable *b* represents the growth ratio, which includes the interest rate as a decimal.

Check for Understanding: Investigation 2

1. D
2. A. Decreasing
   B. Increasing
   C. Increasing
   D. Decreasing
3. 10,000, 15, 1, 5, 4,437
4. It tells you that the rate is given as a decimal. If $r > 0$, then it will be increasing. If it is between 0 and –1, it will be decreasing.

Check for Understanding: Investigation 3

1. D
2. 0.0005
3. $y = 900(0.8)^t$
4. The function is going to be increasing because it has a positive rate. It will have a range of $y \geq P$.

Check for Understanding: Investigation 4

1. C
2. B
3. A. Exponential Growth
   B. Exponential Decay
   C. Exponential Decay
   D. Exponential Growth
4. Look for the first integer coordinate for *x* where Curve A is greater than Curve B. This will occur just after the two curves intersect.

Analyze Exponential Growth and Decay Models *(continued)*

Check for Understanding: Investigation 5

1. C
2. B, E, F
3. 27%
4. The equation can be rewritten as $y = a(1 + 0.00111)^{365t}$. This shows an increase percentage of 0.00111, or 0.111%. The exponent of $365t$ means that this is the percent increase each day, because there are 365 days in a year.

Practice Exercises

1. 63
2. 35, 6
3. 7800
4. 17
5. 4
6. C
7. D
8. 12.0
9. C
10. growth, 3
11. $y = a(1 - 0.05)^x$ B
    $y = a(1 - 0.5)^x$ D, E
    $y = a(1 + 0.5)^x$ C, F
    $y = a(1 + 0.05)^x$ A
12. 16.8

Apply

What Is Happening to Some Animal Populations?

This problem allows students to demonstrate their ability to model and graph an exponential function based on data from a real-world scenario. Students also use the model to make a prediction and determine its reasonableness.

- If students do not have Internet access at home, they may need to watch the video and perform the research at school, or you can provide data for use in the problem.
- You may wish to have a sign-up sheet for different species so there is no overlap. Encourage students to look at both types of populations, increasing and decreasing. For students who are having trouble, suggest keywords to search for such as "declining species" for decreasing populations or "invasive species" for increasing populations.
- As an extension, students can do additional research on their species and give a presentation of their findings to the class.

The Apply problems are rich performance tasks that may have a variety of open-ended solutions. One sample student response for this problem can be found in the Model Lesson.

10.1 Perform Operations on Polynomials

Check for Understanding: Investigation 1

1. A, D, E, F
2. A. III.
   B. II.
   C. I.
3. $\frac{4x^3}{y^3}$
4. The denominator in the rational exponent tells you the number to use for the root in the radical. The denominator of 3 means that the root will be 3, resulting in $\sqrt[3]{3^5x^{10}}$ or $(3x^3)\sqrt[3]{3^2x}$.

Check for Understanding: Investigation 2

1. C
2. $-5x$
3. B
4. Once all of the like terms are combined together, the y-intercept is the constant term of the polynomial. The constant is $4 - (-1) = 5$, so the y-intercept is (0, 5).

Check for Understanding: Investigation 3

1.

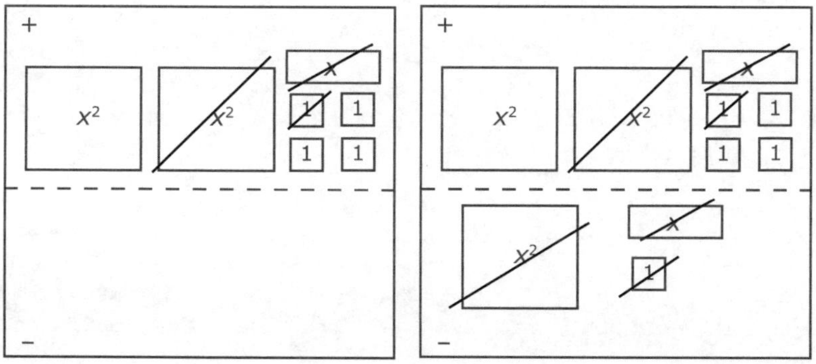

$x^2 + 3$

2. B
3. $-3x^2 + 2x + 0.25$
4. The difference is also a polynomial because after the subtraction of any pairs of like terms, the resulting answer will only contain terms where the variables have whole number exponents.

Perform Operations on Polynomials *(continued)*

Check for Understanding: Investigation 4

1. C
2. $x^2 - 3x + 2$
3.

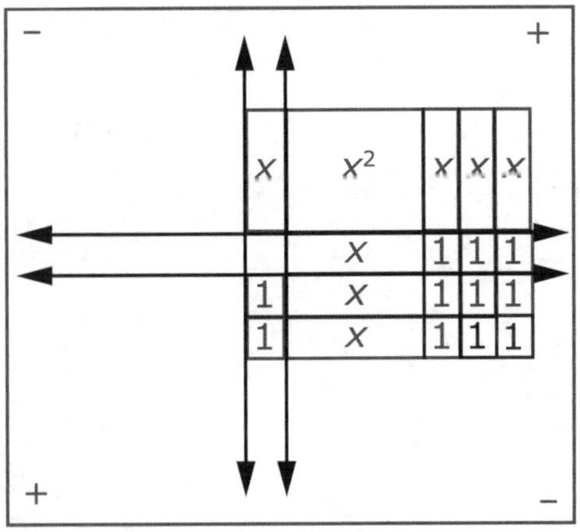

$x^2 + x - 6$

4. If the product has a negative term, then either *a* or *b* are negative or both *a* and *b* are negative.

Check for Understanding: Investigation 5

1. (1,0) and (2,0), (0,12)
2. B
3.

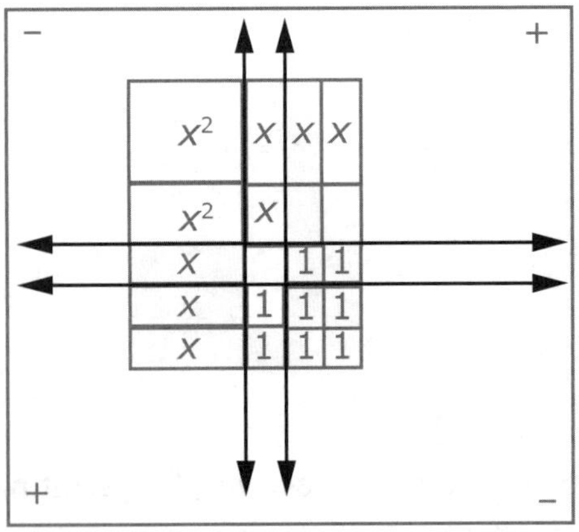

$-2x^2 + 6x - 4$

4. The polynomial will be of degree four. The *x*-intercepts of the polynomial will be the same as the *x*-intercepts of the individual binomials. The *y*-intercept will be the product of the four *y*-intercepts.

Perform Operations on Polynomials *(continued)*

Check for Understanding: Investigation 6

1. B, C
2. A, B, E
3. $x^2 - 64, x^2 + 10x + 25$
4. $$\begin{aligned}(x-a)^2 &= (x-a)(x-a)\\ (x-a)^2 &= x(x-a) - a(x-a)\\ (x-a)^2 &= x^2 - ax - ax + a^2\\ (x-a)^2 &= x^2 - 2ax + a^2\end{aligned}$$

Practice Exercises

1. $12x^5 - 8x^4$
2. C
3. Not a polynomial: C, E, F, H
   Polynomials: A, B, D, G
4. B
5. $6m^4$
6. A; C
7. A, S, S, A, S
8. C
9. D
10. $-3(x+1)^2$ D
    $(3-3x)^2$ C
    $(3x+3)(3x-3)$ A
    $(3x+3)(3x-1)$ B
11. B
12. D

Apply

How Can Polynomials Help Predict Genetics?

In this problem, students explore Punnett squares, diagrams used by biologists to determine possible genetic outcomes from different alleles of a specific gene. They then use a model similar to a Punnett square to show multiplication of binomials, and extend this model to multiply trinomials.

- Remind students to collect like terms when they write out the product. For trinomials, ask them to look for a pattern of where the like terms will be. They should lie on a diagonal of the square.
- As an extension, students can explore other operations such as multiplying a binomial by a trinomial, or dividing a trinomial by a binomial.

The Apply problems are rich performance tasks that may have a variety of open-ended solutions. One sample student response for this problem can be found in the Model Lesson.

10.2 Factor Polynomials

Check for Understanding: Investigation 1

1. B

2.

Expression	Length	Rectangle or Square?	Width
$3x^2 + 5x + 2$	$3x + 2$	rectangle	$x + 1$
$9x^2 + 12x + 4$	$3x + 2$	square	$3x + 2$
$9x^2 - 4$	$3x + 2$	rectangle	$3x - 2$

3.

Expression	Rectangle or Square?	Product of Factors (Length)(Width)
$x^2 + 10x + 25$	square	$(x + 5)(x + 5)$
$x^2 - 12x + 36$	square	$(x - 6)(x - 6)$
$2x^2 + 5x + 3$	rectangle	$(2x + 3)(x + 1)$

4. For the trinomial to be a square, the last term has to be a perfect square. The middle term's coefficient will be twice that square root. If you use 9 and 6, you get that $x^2 + 6x + 9 = (x + 3)(x + 3)$ is a square.

Check for Understanding: Investigation 2

1. B, C

2. A. II.
B. III.
C. I.

3.

Expression	Factored Form	Square Trinomial, Difference of Squares, or Neither
$x^2 + 16x + 64$	$(x + 8)(x + 8)$	square trinomial
$x^2 + 12x - 64$	$(x + 16)(x - 4)$	neither
$x^2 - 16x + 64$	$(x - 8)(x - 8)$	square trinomial
$x^2 - 64$	$(x - 8)(x + 8)$	difference of squares

4. The factored form of $f(x) = \left(x + \frac{p}{2}\right)^2$ means that the x-intercept would be at $\left(-\frac{p}{2}, 0\right)$. This will also be the minimum of the function, so there will be only one intercept.

Factor Polynomials *(continued)*

Check for Understanding: Investigation 3

1.

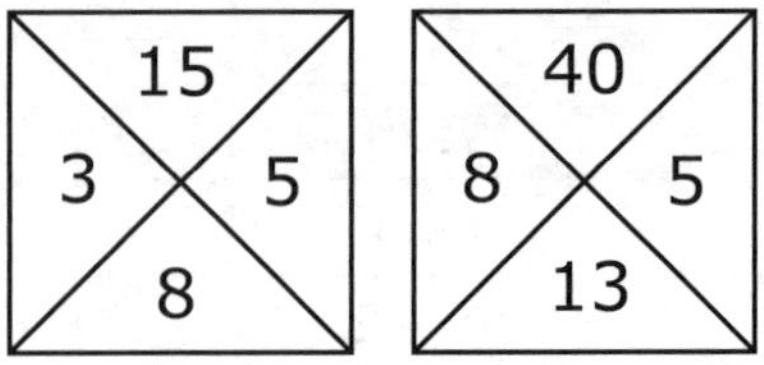

2.

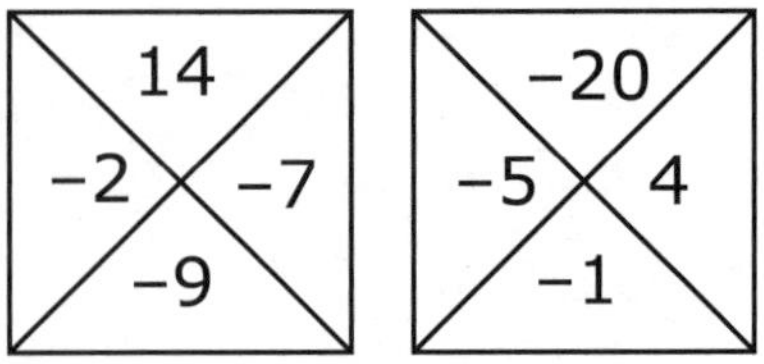

3. B, C

4. The number on top is the product of the numbers on the two sides. The number on the bottom is the sum of the numbers on the two sides.

Check for Understanding: Investigation 4

1.

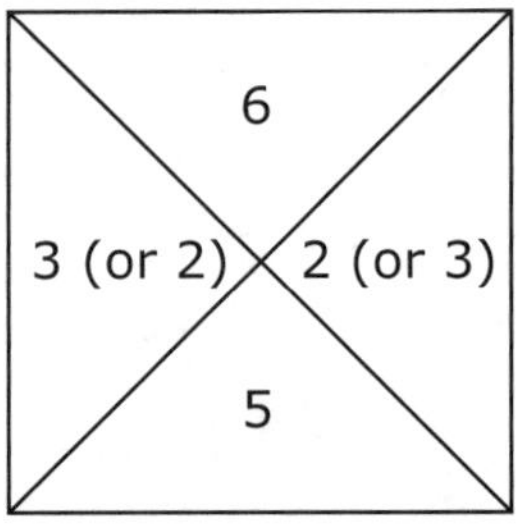

$(x + 3)(x + 2)$

2. A, D

3. Expanded form: $x^2 + x - 12$, Factored form: $(x + 4)(x - 3)$

4. The zeros of the expression are the values of x that cause the expression to equal zero. In factored form, the zeros are the values that make each factor zero. So in this case, $x = -a$ and $x = -b$.

Factor Polynomials *(continued)*

Check for Understanding: Investigation 5

1. C

2. A, D

3. $4(x^2 - 5x + 6), 4(x^2 - 2x - 3x + 6), 4(x(x - 2) - 3(x - 2)), 4(x - 3)(x - 2)$

4. The zeros of the expression can be found from the x values where the graph crosses the x-axis. This means the value of y is zero for these input values of x.

Practice Exercises

1. $(x + 4), (x - 2)$ (either order)

2. B, F, G

3. $2x + 7$

4. 7

5. D

6. A, D

7. $49, x - 7$

8. C

9. D

10. E, I

11. $-6(x^2 - 3x + 12), (-6)(x^2 - 3x + 12),$
$(6)(-x^2 + 3x - 12), (6)(-x^2 + 3x - 12)$

12. $(x - 3)(x - 6)$

Apply

How Can You Predict How Many Diagonals Are in a Polygon?

In this problem, students use a polynomial model to make a prediction of the number of diagonals in an n-sided polygon. They can use any method to discover the model and write the model in both factored and expanded form.

- As an extension, discuss the domain and range of the graph. What are possible numbers of sides of a polygon? What is the meaning of the x-intercept of the graph at $x = 3$?
- You may wish to have students who determined the model in different ways present their methods to the class.
- Students who determine the model by looking at the number of diagonals from each vertex may forget to divide by 2. Ask them to draw the diagonals for polygons with a small number of sides and compare with their model.

The Apply problems are rich performance tasks that may have a variety of open-ended solutions. One sample student response for this problem can be found in the Model Lesson.

11.1 Solve Quadratics

Check for Understanding: Investigation 1

1. 0, 5, −8
2. D, F
3. D
4. B, C
5. The factored form is $2x(x + 10) = 0$. The zero product property means that the linear factors have to equal zero at $x = 0$ and $x = -10$.

Check for Understanding: Investigation 2

1. B, D
2.

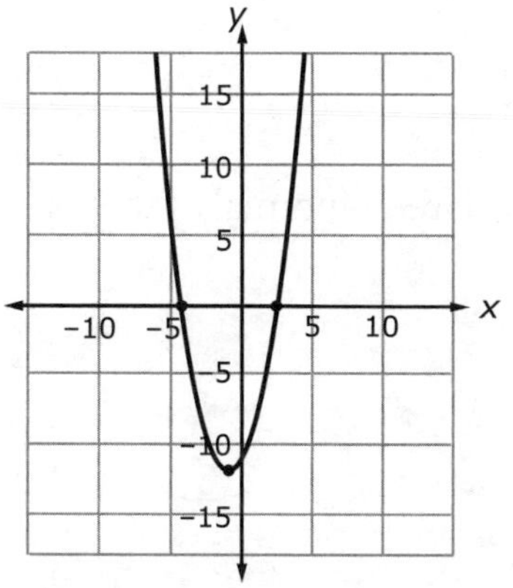

3. D
4. $f(t) = -16t^2 + 96t = -16t(t - 6) = 0 \rightarrow t = 0$ or $t = 6$
$g(t) = -16t^2 + 96t + 432 =$
$g(t) = -16(t^2 - 6t - 27) = -16(t - 9)(t + 3) = 0 \rightarrow t = -3$ or $t = 9$
The second launch stays in the air longer, 9 seconds instead of 6 seconds.

Check for Understanding: Investigation 3

1. B, D
2. F, G
3. $4 - \sqrt{7}$
4. The points of intersection of the two functions represent the x values where $x^2 + 5 = 4x - 7$. If you move everything onto the left side, you get $x^2 - 4x + 12 = 0$, so the points of the intersection are the solutions for x.

Solve Quadratics *(continued)*

Check for Understanding: Investigation 4

1. B

2. A. 3
B. 16, 16
C. +4, 19
D. +4, $\pm\sqrt{19}$
E. $-4 \pm \sqrt{19}$

3. A, B

4. You want to use completing the square if you cannot factor the binomial. So if c has two factors that add up to b, then you can factor instead of completing the square.

Check for Understanding: Investigation 5

1. D

2. $x = 2$

3. B

4. A double root means that the factored form of the quadratic equation is a squared binomial of the form $y = (x - a)^2$. This expands into a quadratic form $y = (x - a)^2 = x^2 - 2ax + a^2$.

Practice Exercises

1. D

2. B

3. $2\sqrt{6}, -2\sqrt{6}$ (either order)

4. −2, 6 (either order)

5. $6, \frac{2}{3}$ (either order)

6. B

7. D

8. E

9. 1.25, 6.25

10. D, C

11. 0.5, 0 (either order)

12. C, E

13. E

Apply

A Very Special Number

This problem provides students opportunities to apply what they have learned about solving quadratic equations to investigate a ratio.

- Do not tell students that this is phi. Rather than looking it up to find its value, they should find its value through their research.
- If, however, research is not an option, you may wish to tell them that the numbers are called the Golden Ratio. The Golden Ratio appears in nature the most, in the arrangement of flower petals, sunflower seeds, and pinecone seeds, and is the ratio thought to be the most beautiful in art and architecture. This number is called phi.

The Apply problems are rich performance tasks that may have a variety of open-ended solutions. One sample student response for this problem can be found in the Model Lesson.

11.2 Analyze Quadratic Equations

Check for Understanding: Investigation 1

1. Divide each term by the lead coefficient. = **Step 2:** $x^2 + \frac{b}{a}x = -\frac{c}{a}$

   Take the square root of both sides. = **Step 5:** $x + \frac{b}{2a} = \pm\sqrt{\frac{b^2 - 4ac}{4a^2}}$

   Subtract the constant from both sides. = **Step 1:** $ax^2 + bx = -c$

   Move all terms to the right and simplify the square root. = **Step 6:** $x = -\frac{b}{2a} \pm \frac{\sqrt{b^2 - 4ac}}{2a}$

   Add the square of half of the linear term to both sides. = **Step 3:** $x^2 + \frac{b}{a}x + \left(\frac{b}{2a}\right)^2 = -\frac{c}{a} + \left(\frac{b}{2a}\right)^2$

   Write the squared binomial and simplify the right side. = **Step 4:** $\left(x + \frac{b}{2a}\right)^2 = \frac{b^2 - 4ac}{4a^2}$

2. A

3. If the number in the radical is negative, there will not be a real root. This will happen anytime $4ac > b^2$, such as in $20x^2 - 2x + 100 = 0$.

Check for Understanding: Investigation 2

1. B
2. A
3. C, D, E
4. D
5. If $D > 0$ there are two roots. If D is a perfect square, there will be two rational roots. If D is not a perfect square, the two roots are irrational. If $D = 0$, there will be one rational root. If $D < 0$, there are no real roots.

Check for Understanding: Investigation 3

1. B
2. 64
3. 3.75
4. The axis of symmetry can be found by the term $-\frac{b}{2a}$ in the quadratic equation. The maximum or minimum of the parabola will be at this value, so evaluate $f\left(-\frac{b}{2a}\right)$ to find it.

Analyze Quadratic Equations *(continued)*

Practice Exercises

1. A, C, D

2. 4, 10, −3

3. B

4. 1152, 2

5. No real solutions: A, F
Exactly one real solution: B, E
Exactly two real solutions: C, D

6. 0, 25

7. C

8. A

9. A, C, D

10. −32, 0

11. 5

12. B, C

Apply

How Much Would You Charge for Your App?

This problem allows students to apply quadratic functions to solve a real-world problem. Students may solve the problem either graphically or algebraically.

- If you use this problem as a homework assignment, you may want to set up the problem during class time by having students watch the video App Development.
- If research is not an option, you may wish to have students estimate the number of people in their target market based on what they already know.

The Apply problems are rich performance tasks that may have a variety of open-ended solutions. One sample student response for this problem can be found in the Model Lesson.

12.1 Analyze Graphs of Quadratic Functions

Check for Understanding: Investigation 1

1. D

2. B

3. A, E

4. A. 4
B. 120
C. $2,880

5. The constant term in the standard form is the y-intercept and this is the income from last year's play. The vertex form gives the coordinates for the vertex, which are the maximum possible income, y, based on the number of $4 ticket price increases, x.

Check for Understanding: Investigation 2

1. A. upward, $x = 10$
B. minimum, –8, 10
C. (0, 192), (8, 0), (12, 0)

2. A. VII.
B. V.
C. I.
D. VII.
E. V.

3. 0.01, 10, 1
(or $y = \frac{1}{100}(x - 10)^2 + 1$)

4. The general vertex form is $y = a(x - h)^2 + k$. Substitute the x value from the vertex for h the y value for k. Then to calculate a, substitute 0 for x in the equation and the y-intercept for y, and solve for a.

Check for Understanding: Investigation 3

1. A. 8, 8, 16, 16, 4, –4
B. 5, 6, 5, 6, 9, 45, 5, 3, 4
C. –0.5, –12, –0.5, 12, 36, –18, –0.5, 6, 22

2. C

3. A. $w(400 - 2w)$ (or $400w - 2w^2$ or $-2w^2 + 400w$)
B. $-2(w - 100)^2 + 20{,}000$
C. 20,000, 100, 200

4. Since the vertex is (h, k), a reasonable domain and range are positive. The domain includes all values that are h units on either side of the x value of the vertex, so it must be $(0, 2h)$ (or $0 < w < 2h$) and the range, the values for the length, must be $(0, k]$ (or $0 < l \leq k$).

Analyze Graphs of Quadratic Functions *(continued)*

Check for Understanding: Investigation 4

1. -0.75, 90, (or $-\frac{3}{4}$, 90)
2. A. $-0.75x^2 + 90x$ (or $-\frac{3}{4}x^2 + 90x$)
   B. $-0.75(x - 60)^2 + 2700$ (or $-\frac{3}{4}(x - 60)^2 + 2700$)
   C. \$2,688
   D. 42
3. A. (0, 0), (0, 0), (120, 0)
   B. (60, 2,700)
   C. \$2,700, \$60, 45
   D. 120, 2,700
4. Answers will vary. Sample Response: Even though placing 45 dogs will maximize income from fees, the purpose of the group is to find homes for the dogs, so they should place the maximum number of dogs they can, while covering their expenses.

Check for Understanding: Investigation 5

1. A. V.
   B. I., III., VI.
   C. II., IV.
2. A. 1.8, 0.95
   B. 0.66
   C. $H(t) = -16.8t^2 + 11.09t + 0$
   D. 11.09
3. $g(x) = -\frac{5}{4}x^2 + 10x - 9$
4. $h(x) = 3x^2 - 6x - 2$
5. Use the time between bounces to write the factored form of the quadratic model. Measure in feet and use -16.8 for a. Use the midpoint between the two times to substitute in the model for the maximum height. Then set the first time measure to 0 and change the second by the same amount to write the equation in standard form, and the coefficient of the linear term gives the initial velocity.

Analyze Graphs of Quadratic Functions *(continued)*

Practice Exercises

1. Parabola that opens up: B, C
   Parabola that opens down: A, F
   Not a parabola: D, E
2. A, C, E
3. −6, 0
4. D
5. −3, −25
6. Standard form: B, C
   Vertex form: A, D, E
7. −2, −4
   −2, −2
8. B, C, E
9. 7
   40
10. D
    C
11. B

Apply

How Can Math Help You Create a Successful Business?

In this problem, students analyze the relationship of a demand function and a revenue function. They interpret the key features of the graphs, and determine the price of the item that will maximize revenue.

- Students who do not have Internet access at home may need to watch the video at school.
- The demand function gives the number of items that would be sold at a given price, x. Discuss the demand function with students. Why is it linear? Why is it decreasing?
- As an extension, students can explore what price of the item will produce a maximum profit. Have them estimate a cost per item and then graph the cost function. Look at the region where the revenue is greater than the cost; this is where the profit is positive. The x value in this region where the difference is greatest represents the price for the maximum profit.

The Apply problems are rich performance tasks that may have a variety of open-ended solutions. One sample student response for this problem can be found in the Model Lesson.

13.1 Explore Recursive Functions

Check for Understanding: Investigation 1

1. A, C, D
2. $d = 6, a_1 = -10, a_2 = -4$
3. A.

Term	(a_0)	a_1	a_2	a_3
Payment	0	1	2	3
Remaining Balance	425	380	335	290

B. $a_n = 380 - 45(n - 1)$ or $y = 425 - 45x$

4. D
5. Similarities: In the explicit arithmetic formula, the rate of change and the initial starting value represent the slope and the y-intercept of the related linear equation.

   Differences: The linear equation represents a continuous line with a domain of all real numbers but the arithmetic sequence is a discrete set of data that is not continuous. For example, in the, payment of the snowboard, there is no data between the monthly payment, she is only paying one payment a month at one time.

Check for Understanding: Investigation 2

1. 7

Row	A	B	. . .	P
Number of Seats	58	65	. . .	163

2. A
3. 24
4. If an arithmetic series of numbers is expanded as a sum of terms and the reverse expansion of the sum is written underneath the first expansion, the individual sums of each column of terms are the same value as the sum of the first term and the last term. Therefore, the sum of any number of terms can be found by adding the first term to the last term, then multiplying by the number of terms, but the resulting product value would be twice the sum, so we need to divide by 2.

Check for Understanding: Investigation 3

1. A. $-\frac{1}{2}$, B. none, C. $\frac{2}{3}$, D. none
2. $r = -4, a_4 = -12$
3. Growth rate: 1.5, Absences expected on closing day: 615
4. When $r < -1$, the terms will oscillate between positive and negative values no matter what the sign of a_1 is. If $r > 1$, the terms will all stay the same sign, (if a_1 is negative when $r > 1$, all the terms will remain negative). When $-1 < r < 1$, each successive term in the sequence gets closer to zero but when $r < -1$, the terms get further from zero and oscillate between positive and negative as the term position increases.

Explore Recursive Functions *(continued)*

Check for Understanding: Investigation 4

1. B, D
2. Month: 13, Number of new contracts expected: 8
3. C
4. Using an expanded sum of 4 terms: $2 \cdot S_4 = 6 + 12 + 24 + 48$. If I multiply the sum through by the common ratio, all the terms shift down 1: $2 \cdot S_4 = 6 + 12 + 24 + 48$. When subtracting the two sums I get the first term minus one past the last term: $2 \cdot u_4 = 48$. In subtracting the summation notation: $(a_1 - ra_n)$. Solving for S_4 I get: $S_4 = \frac{(a_1 - ra_n)}{(1 - r)}$.

Check for Understanding: Investigation 5

1. A. $\frac{5}{2}$, Divergent, none
   B. $-\frac{2}{3}$, Convergent, −16.2
   C. $\frac{3}{2}$, Divergent, none
   D. $\frac{1}{10}$, Convergent, $\frac{1}{9}$
2. Common Ratio: $\frac{3}{5}$ or 0.6, Sum: 150 ml
3. When using the partial sum formula for an infinite geometric series $S_n = \frac{a_1 - a_n \cdot r}{1 - r}$ where r is a value between −1 and 1, the series is convergent. This means the last term approaches zero: $a_n \to 0$. Zero can be substituted into the partial sum so the numerator reduces to only a_1, resulting in an infinite sum: $S = \frac{a_1 - 0 \cdot r}{1 - r} = \frac{a_1}{1 - r}$.

Explore Recursive Functions *(continued)*

Practice Exercises

1. D

2. 350

3. $r = -0.5$
$a_2 = -72$
$a_3 = -36$

4. −12,228

5. B, C, D

6. B, D, E, I

7. D

8. A, D

9. B

10. 98

Apply

How Many Digits Are Needed in a Telephone Number?

This problem allows students to explore the pattern between the number of digits in a phone number and the number of possible phone numbers. Students determine how many digits a phone number must have so that every person on the planet can have a different phone number. This process is repeated with different parameters.

The Apply problems are rich performance tasks that may have a variety of open-ended solutions. One sample student response for this problem can be found in the Model Lesson.

13.2 Explore Inverse Functions

Check for Understanding: Investigation 1

1. D
2. the same, different, is not, is, is
3. The graph that Elon drew can be represented by the equation $\frac{1}{2}(x+8) = |y|$, and this relation is not a function for the reason that Elon gave. However, the graph that Elon drew is not correct, because a relation represented by a table is discrete, meaning that when it is graphed, lines cannot be drawn through the plotted points. Since each *x* value in the table is associated with one and only one *y* value, the relation represented by the table is a function.

Check for Understanding: Investigation 2

1. A, C, G
2. A. $6 - x$
   B. $3 - x$
   C. $-36 + 24x - 4x^2$
   D. $\frac{x}{4}$
   E. $-6 + 4x$
   F. $6 + 2x^2$
3. The first error she made was in the second line when reducing the second term: $\frac{2x+10}{2}$ reduces to $x + 5$. Then Jayden didn't distribute the minus sign in front of the second term. This subtraction is rewritten as a sum of a product: $4 + -1(x+5)$. The -1 must be distributed throughout to get: $4 - x - 5$ which can be rewritten as $k(p(x) = -x - 1$. Jayden needs to remember to reduce factors within terms and distribute negative values.

Check for Understanding: Investigation 3

1. 9
2. Step 1: Add 10, Step 2: Divide by 3, Step 3: Subtract 4
3. $3(n+4) - 10$
4. $\frac{(r+10)}{3} - 4$
5. When two functions form a composite function, the domain of the first function is computed to its range value, and the set of range values from the first function becomes the set of domain values for the second function. As these new domain values are computed through the second function, their resulting range values may or may not be related to the original domain of the first function. When two functions are inverses of each other, the domain values of the first function result in range values that become domain values for the second, inverse function. When these new domain values are computed in this inverse function, the resulting range values must match the original domain values.

Explore Inverse Functions *(continued)*

Check for Understanding: Investigation 4

1. A.

x	$f^{-1}(x)$
–1	9
0	6
1	3
2	0
3	–3

B.

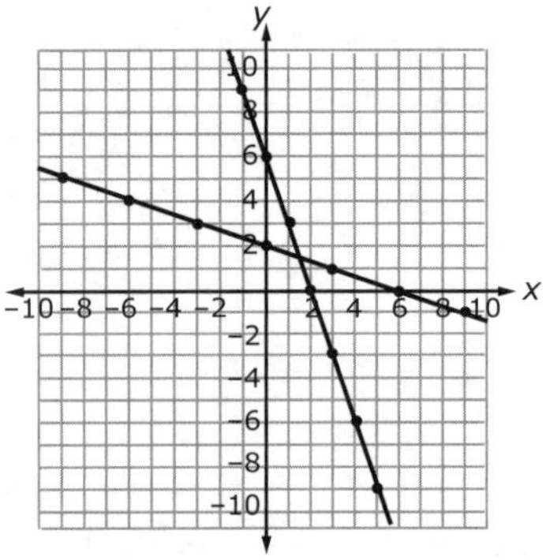

2. $f^{-1}(x) = -3x + 6$
3. All inverses reflect over the line $y = x$ because a property of inverse functions is that each member of the domain maps to a range in the original function. The inverse function must use each member of the original range as a domain value, an input for the inverse function resulting in a range that matches the original domain.

   This process results in $g^{-1}(x) = \frac{(x+8)}{4}$. This is not simplifying; this is solving for the y variable. Ty made algebraic errors in his solving:

   $x = -4y - 8;\ x + 8 = -4y;\ \frac{(x+8)}{-4} = y$ so $g^{-1}(x) = -\frac{x}{4} - 2;\ -\frac{x}{4} - 2 = y$

Check for Understanding: Investigation 5

1. C, D
2. $T(a) = 5a - 317$ where T: Test score and a: average goal
3. 83%
4. 98%
5. The fifth test score cannot be below 0%, and there is no extra credit so the highest score he can earn is 100%. Domain for $A(t)$ is [0, 100]. Since Darin already has four scores, the lowest his average will be with a score of 0% will be 63%. The highest average, if he earns 100% will be 84%. Range of $A(t)$ is [63, 84].

   The inverse function then will have a domain such that the only realistic average percentages obtainable are between 63% and 84%, knowing that the related score for the fifth test must fall between 0% and 100%.

Explore Inverse Functions *(continued)*

Practice Exercises

1. A
2. C, D
3. B
4. B
5. A, F
6. C
   B
   A
7. 0.40, −2
8. −8
9.

t	$f^{-1}(t)$
50	10
122	50
95	35

Apply

Does How Fast You Drive Affect How Much Fuel You Use?

In this problem, students begin by watching a video about race car performance. They are given a table of values that relates one race car's gas mileage to its speed. Students model the data with a function and identify a reasonable domain for the function. They then graph the inverse of the function and explain what the inverse represents. Finally, students use the function and its inverse to predict the speed at which the car's gas mileage is the greatest and to estimate how much fuel the car would use during a 500-mile NASCAR race. Note that a typical car in a NASCAR race averages about 170 miles per hour; you may wish to provide students with this information, or they can research it themselves by using the Internet.

The Apply problems are rich performance tasks that may have a variety of open-ended solutions. One sample student response for this problem can be found in the Model Lesson.

13.3 Explore Function Transformations

Check for Understanding: Investigation 1

1. Equation: $y = 15x + 25$
   Graph: C
   Dilation: none
   Translation: vertical up 25
2. Equation: $y = 18x + 25$
   Graph: A
   Dilation: multiplied by 1.2
   Translation: vertical up 25
3. Equation: $y = 12x + 30$
   Graph: B
   Dilation: multiplied by 0.8
   Translation: vertical up 30
4. Answers will vary. Amy sells her handmade scarves for $28 a piece. To sell them at the county fair, she will increase her price to $35 because she has to pay $120 for the fair booth she is using. $y = 35x–120$. (Be sure to also assess students' graphs.)

Check for Understanding: Investigation 2

1.

$f(x) = x^2 - 6$	$g(x) = \lvert x - 6\rvert$	$h(x) = \sqrt{x} + 6$	$k(x) = \sqrt{x + 6}$	$p(x) = (x - 6)^2$
down 6	right 6	up 6	left 6	right 6

2.

$f(x) = -x^2$	$f(x) = (-x)^2$	$p(x) = -\lvert x\rvert$	$g(x) = \lvert -x\rvert$	$h(x) = -\sqrt{x}$	$k(x) = \sqrt{-x}$
over x-axis	over y-axis	over x-axis	over y-axis	over x-axis	over y-axis

3.

$f(x) = 4x^2$	$g(x) = \lvert 5x\rvert$	$h(x) = 0.5\sqrt{x}$	$k(x) = \sqrt{0.5x}$	$p(x) = 0.5\lvert x\rvert$
vertical stretch	horizontal compression	vertical compression	horizontal stretch	vertical compression
$q(x) = \left(\frac{1}{4}x\right)^2$	$g(x) = \frac{3}{2}\lvert x\rvert$	$t(x) = \frac{1}{4}x^2$	$h(x) = \sqrt{\frac{3x}{2}}$	$f(x) = (4x)^2$
horizontal stretch	vertical stretch	vertical compression	horizontal compression	horizontal compression

4. When a function has a vertical stretch, it will have the same effect as a horizontal compression, but by a different factor because of when the multiplication by the factor takes place.

   For example, given $g(x) = 4\sqrt{x}$, every domain value will first have a square root taken, then the output of the square root function is multiplied by 4. This is a composite function where the radical is computed first, then multiplied. Comparing $g(x) = 4\sqrt{x}$ to $f(x) - \sqrt{4x} = \sqrt{4}\cdot\sqrt{x} = 2\sqrt{x}$, we can see that through simplification of the function, $f(x) = \sqrt{4x}$ will cause a stretch, but not as severe as $g(x) = 4\sqrt{x}$. In the same way, a horizontal stretch will have the same effect as a vertical compression, but by a different factor.

Explore Function Transformations *(continued)*

Check for Understanding: Investigation 3

1.

Parent Function	Vertical Translation	Horizontal Translation	Dilation	Reflection
$f(x) = x^2$	Up 2	None	Stretch: 3	Over x-axis

2.

Parent Function	Vertical Translation	Horizontal Translation	Dilation	Reflection
$g(x) = \|x\|$	Down 6	Left 4	Compress: 0.5	Over y-axis

3. $h(x) = -1.5\sqrt{-x+2} - 4$

4. $k(x) = -2|x+5| - 3$

5. When a function has a negative coefficient as a factor outside of the parent function rule, the output of the parent function changes sign after the parent function rule making the output change in sign as a range value, therefore reflecting it over the x-axis. When the negative coefficient is acting on x before the parent function rule is applied, the x value changes sign so the resulting graph will appear reflected over the y-axis.

Practice Exercises

1. C

2. B

3. A, B, D

4. (a) $f(x) = \sqrt{-\frac{1}{2}x}$
(b) $f(x) = \sqrt{-2x}$
(c) $f(x) = \sqrt{2x}$
(d) $f(x) = \sqrt{x}$
(e) $f(x) = -\sqrt{2x}$

5. −1, 3

6. A, D

7. C

8. B

9. C

10. (5, 1)

Apply

Can You Launch the Bird through the Hoop?

This problem allows students to analyze a game situation involving functions and transformations. Students will write equations to satisfy various conditions for the bird and the hoop, describe the attributes of each equation, and compare the equations written to the original equation.

The Apply problems are rich performance tasks that may have a variety of open-ended solutions. One sample student response for this problem can be found in the Model Lesson.

14.1 Operate with Polynomials

Check for Understanding: Investigation 1

1. A. IV.
   B. V.
   C. VI.
   D. III.
   E. II
   F. I.
2. A, E
3. 3, trinomial
4. B, C, F
5. The volume of the first container is x^3 cubic feet, while the volume of the second container in cubic feet can be represented by the expression $(x - 5)(x - 2)(x - 3)$. Multiplying the factors and combining like terms gives $x^3 - 10x^2 + 31x - 30$. The difference in the volumes can be represented by the expression $x^3 - (x^3 - 10x^2 + 31x - 30)$, so the first container's volume is $10x^2 - 31x + 30$ cubic feet greater than that of the second container.

Check for Understanding: Investigation 2

1. B
2. B, F
3.

Equation	Number of Rational Roots	Number of Irrational Roots	Number of Non-Real Roots
$x^4 - 256 = 0$	2	0	2
$x^4 + 11x^2 + 18 = 0$	0	0	4
$x^4 - 10x^2 + 24 = 0$	2	2	0
$x^4 - 25 = 0$	0	2	2
$x^4 - 50x^2 + 49 = 0$	4	0	0
$x^4 - 13x^2 + 30 = 0$	0	4	0

4. Yes, it was. First factoring out $\frac{1}{12}x^2$ to get $d(x) = \frac{1}{12}x^2(x^2 - 12x + 36)$ and then factoring again to get $d(x) = \frac{1}{12}x^2(x - 6)(x - 6)$ shows that the zeros of the function are $x = 0$ and $x = 6$. Therefore, the stock was at its opening price 6 hours after the open, or 6 hours after 9:30 a.m., which is 3:30 p.m.

Operate with Polynomials *(continued)*

Check for Understanding: Investigation 3

1. B, F
2. 1, 1, 1
3. A, D
4. C
5. Neither person made a mistake. The reason why they didn't get the same result is because Lisa's result is not fully factored. There is not an easy way to factor $x^4 + 4x^2 + 16$, but it is possible to verify that it can be factored into $(x^2 - 2x + 4)(x^2 + 2x + 4)$ by multiplying these two factors together. Distributing gives $x^4 - 2x^3 + 4x^2 + 2x^3 - 4x^2 + 8x + 4x^2 - 8x + 16$, and combining like terms gives $x^4 + 4x^2 + 16$, so Lisa could have factored more.

Check for Understanding: Investigation 4

1. C, D
2. third, fifth, second, fourth, sixth
3.

Binomial Expression	Expanded Form
$(4x^2 + 3y)^3$	$64x^6 + 144x^4y + 108x^2y^2 + 27y^3$
$(2x - y^3)^4$	$16x^4 - 32x^3y^3 + 24x^2y^6 - 8xy^9 + y^{12}$
$(3x^4 - 2y)^3$	$27x^{12} - 54x^8y + 36x^4y^2 - 8y^3$
$(x^3 + 5y)^4$	$x^{12} + 20x^9y + 150x^6y^2 + 500x^3y^3 + 625y^4$
$(5x - 4y^2)^3$	$125x^3 - 300x^2y^2 + 240xy^4 - 64y^6$

4. First, Joe went to the wrong row in Pascal's Triangle. He should have gone to the row with seven, and not six, numbers in it. The fourth number from the left in this row is 20, and not 10. Joe also raised each of the terms in the binomial expression to the wrong power. He should have multiplied 20 by the product of x^{6-3} and $(-2y^2)^3$, and not the product of x^{6-4} and $(-2y^2)^4$, to get $(20)(x^{6-3})(-2y^2)^3 = (20)(x^3)(-8y^6) = -160x^3 y^6$ for the fourth term.

Operate with Polynomials *(continued)*

Practice Exercises

1. $3x^2 + 2$
2. A
3. C
4. C
5. B
6. D
7. D
8. $x + 2, x + 3$
9. True: B, C, D
   False: A, E, F
10. $20x^3y^3$
11. $3x + 2$
12. $36t^2 - 9n^4$

Apply

Have You Seen the Lizards?

This problem allows students to demonstrate their ability to factor a fifth degree polynomial function and recognize the real-world meaning of the zeros. Students graph the function, determine the domain and range in the real-world context, and use the function to make a prediction.

- Students could graph the function by hand by making a table of values, but in the interest of time and to avoid busy work, you may wish to allow students to use the Graphing Calculator to graph and analyze the function.
- If you use this problem as a homework assignment, you may want to set up the problem during class time by having students watch the video, Where Did All the Lizards Go?

The Apply problems are rich performance tasks that may have a variety of open-ended solutions. One sample student response for this problem can be found in the Model Lesson

14.2 Explore Polynomial Factors

Check for Understanding: Investigation 1

1. B, E
2. divisor or quotient, divisor or quotient, dividend
3. A. V.
   B. I.
   C. VI.
   D. II.
   E. IV.
   F. III.
4. does not, is
5. Yes, it is possible to divide a polynomial by a polynomial and get a negative remainder, so Madeline may not have made a mistake. For example, suppose she divided $18x^2 + 19x - 13$ by $2x + 3$. If she did not make any errors, she should have gotten a quotient of $9x - 4$ with a remainder of -1.

Check for Understanding: Investigation 2

1. B, E
2. C
3. $-2, \frac{3-\sqrt{5}}{2}, 2, \frac{3+\sqrt{5}}{2}$
4. Since the equation has roots of $x = \pm 2i$, you can multiply $x - 2i$ and $x + 2i$ to get $(x - 2i)(x + 2i) = x^2 + 2ix - 2ix - 4i^2 = x^2 - 4i^2 = x^2 - (-4) = x^2 + 4$. Then you can use long division to find the quotient of $4x^4 + 11x^3 + 13x^2 + 44x - 12$ and $x^2 + 4$, which is $4x^2 + 11x - 3$. Next, you can factor $4x^2 + 11x - 3$ to get $(4x - 1)(x + 3)$. Finally, you can set each factor equal to 0 and solve for x to get $x = \frac{1}{4}$ and $x = -3$, so these are the other roots of the equation.

Check for Understanding: Investigation 3

1. B, F
2. D
3. $x^3 - 64$, $x - 4$, $x^2 + 4x + 16$, difference, cubes
4. It is not possible to do directly, because synthetic division only works when you are dividing by a linear factor. However, you could factor $x^2 - 9$ as $(x - 3)(x + 3)$. Then you could use synthetic division to divide the polynomial $x^3 + 6x^2 - 9x - 54$ by $x - 3$, and after that, you could divide the remaining factor by $x + 3$. The remainder should be 0 both times, showing that both $x - 3$ and $x + 3$, and their product, $x^2 - 9$, are factors of $x^3 + 6x^2 - 9x - 54$.

Explore Polynomial Factors *(continued)*

Check for Understanding: Investigation 4

1. 16
2. B, C, F
3. D
4. According to the rational root theorem, the possible rational roots of the equation are $x = \pm 1, \pm 3, \pm 5, \pm 15, \pm\frac{1}{2}, \pm\frac{3}{2}, \pm\frac{5}{2}, \pm\frac{15}{2}, \pm\frac{1}{3}, \pm\frac{5}{3}, \pm\frac{1}{6}$, and $\pm\frac{5}{6}$. However, none of these are roots of the equation, because the left side can be factored so that the equation becomes $(6x^2 + 5)(x^2 - 3) = 0$, and setting neither factor equal to 0 and solving for x gives rational roots. The equation actually has two non-real, complex roots and two irrational roots.

Check for Understanding: Investigation 5

1. E, F, H
2. D
3. $0, \frac{1}{2}, i\sqrt{2}$
4. Edwin listed all of the possible rational roots of the equation correctly and performed the synthetic division correctly when identifying only $x = -3$ as a rational root of the equation. His mistake was that he forgot that $x = -3$ could be a double root. If he would have divided $8x^3 + 24x^2 + x + 3$ by $x + 3$ using synthetic division, he would have gotten a remainder of 0 and found that the fully factored form of the equation is $(x + 3)(x + 3)(8x^2 + 1) = 0$.

Practice Exercises

1. A, C, D
2. C
B, E
3. E
4. $2x^3 - x^2 - 27x + 36$
5. A, B, C, E
G, I
6. C
7. Part I: 2, −5, 1
Part 2: $-\sqrt{5}, 0$
$\sqrt{5}, 0$
$-1, 0$
8. B, F, C, D
9. H, J, Q
10. $(x + 2)(x + 2)(x - 2)(x + 3)$

Apply

Will You Go with the Flow?

This problem provides students with opportunities to expand their methods for dividing polynomials by learning Paolo Ruffini's synthetic division and modeling the algorithm with a flowchart. Students use the flowchart to solve several problems. Ruffini's synthetic is a special case of synthetic division. It can be used when the divisor is a linear factor. If students will not be able to research the method, use the Synthetic Flowchart, which is shown in the sample student response.

The Apply problems are rich performance tasks that may have a variety of open-ended solutions. One sample student response for this problem can be found in the Model Lesson.

14.3 Analyze Polynomial Functions

Check for Understanding: Investigation 1

1. C, D, E
2. odd, negative
3. ∞, ∞
4. Theo's reasoning is not correct. A positive or negative average rate of change within an interval does not mean that the rate of change is always positive or negative within that interval. If Theo graphed the function, he would see that in the interval $[2.5, 3.5]$, the function's rate of change is both positive and negative, but more positive than negative, so the average rate of change is positive. The function actually has a minimum at about $x = 2.9$.

Check for Understanding: Investigation 2

1. A, E, F
2. C
3. 4, –2, (0,–16), $x^3 - 12x - 16$
4. If a quartic function with a positive leading coefficient has two zeros, each with a multiplicity of 2, its graph will decrease from left to right until it bounces off the x-axis and then go back down until it bounces off the x-axis again. This will produce a positive local maximum between the two zeros. No other multiplicities will result in this behavior, so the only equation Barbara could write is $y = (x + 5)^2(x - 3)^2 = x^4 + 4x^3 - 26$ $x^2 - 60x + 225$.

Check for Understanding: Investigation 3

1. A
2. A, C, D
3. If A represents the length, and B represents the width, the function can be written as $y = (x)(A - 2x)(B - 2x)$. The graph shows that the function's zeros are $x = 0$, $x = 18$, and $x = 30$. The first factor is equal to 0 when $x = 0$. For the second factor to be equal to 0 when $x = 30$, A would have to equal 60. For the third factor to be equal to 0 when $x = 18$, B would have to equal 36. Therefore, the dimension of the piece of cardboard are 60 cm by 36 cm.

Analyze Polynomial Functions *(continued)*

Check for Understanding: Investigation 4

1. B

2.

Function	End Behavior as $x \to -\infty$	End Behavior as $x \to +\infty$
$g(x) = x^5 + x^4 - 9x - 9$	$g(x) \to -\infty$	$g(x) \to +\infty$
$h(x) = -x^4 - 2x^3 + 3x^2 + 8x + 4$	$h(x) \to -\infty$	$h(x) \to -\infty$
$k(x) = x^6 - 3x^5 + x - 5$	$k(x) \to +\infty$	$k(x) \to +\infty$
$m(x) = -x^5 + 5x^4 - 2x + 10$	$m(x) \to +\infty$	$m(x) \to -\infty$
$n(x) = x^4 + 9x^3 + 14x^2 - x - 7$	$n(x) \to +\infty$	$n(x) \to +\infty$
$p(x) = -x^6 + 6x^5 - x^2 + 8x - 12$	$p(x) \to -\infty$	$p(x) \to -\infty$

3. D, H, J

4. Norah is right. First, you could determine the possible rational roots of the equation $x^4 - 2x^3 - 19x^2 + 32x + 48 = 0$. They are $x = \pm1, \pm2, \pm3, \pm4, \pm6, \pm8, \pm12, \pm16, \pm24$, and ±48. By testing these, you could determine that the factored form of the function is $r(x) = (x + 1)(x - 3)(x - 4)(x + 4)$. This means the function's graph crosses the x-axis four times. Since $r(x) \to +\infty$ as $x \to +\infty$, the function's graph must change direction exactly three times.

Analyze Polynomial Functions *(continued)*

Practice Exercises

1. D

2. B

3. C, B
D, A
C, D

4. A. $p(x) = -4x^4 + 4x^2 - 3$
B. $p(x) = 3x^2 - 7$
C. $p(x) = -3x^3 + 2x^2 - x - 1$
D. $p(x) = 2x^5 + 8$

5. B

6. B, C

7. A

8. D, E
A, B
C, F

9. −1
−1
2
3

Apply

How Did the Function Cross the Bridge?

This problem provides students with opportunities to apply what they have learned about polynomial and piecewise functions to create a model for an unusual curve.

- Students will need the Graphing Calculator as they define a piecewise function to model the bridge.
- You may wish to allow students to compare their piecewise functions and discuss which is the best model.

The Apply problems are rich performance tasks that may have a variety of open-ended solutions. One sample student response for this problem can be found in the Model Lesson.

15.1 Develop Rational Expressions

Check for Understanding: Investigation 1

1. A. V.
   B. III.
   C. VI.
   D. I.
   E. II.
   F. IV.
2. A, F
3. To simplify the expression, first the numerators and denominators in the original expression should both be factored to get $\frac{(x-y)^2}{-y(1-x)} \div \frac{5(x-y)}{x(1-x)}$, so $x \neq 0, y \neq 0$, and $x \neq 1$. Next, the expression should be written as the product of two fractions so that it becomes $\frac{(x-y)^2}{-y(1-x)} \cdot \frac{x(1-x)}{5(x-y)}$, so $x \neq y$. Finally, factors should be cancelled out and the two fractions should be multiplied to get the simplified form of the expression, which is $-\frac{x(x-y)}{5y}$.

Check for Understanding: Investigation 2

1. D
2. B, F, H
3. In the third line of his work, when Gabe wrote each of the fractions in the expression with a common denominator, he multiplied the numerator of both the first fraction and the third fraction by the wrong factor. He should have multiplied the numerator of the first fraction by $n+4$ and the numerator of the third fraction by $n-4$. He got lucky in this case. With another rational expression, he may not have gotten the right answer.

Check for Understanding: Investigation 3

1. A
2. $5x-1$ or $-1+5x$
3. Jonah did not make a mistake when simplifying the original expression. When substituting for x in both the original expression and the simplified expression, he needs to choose a value that does not cause the denominator of any of the fractions within the original expression to be 0. For example, had he substituted 2 for x, he would have gotten $\frac{2}{3}$ for both expressions.

Check for Understanding: Investigation 4

1. A. V.
   B. IV.
   C. VIII.
   D. VI.
   E. III.
   F. VII.
   G. II.
   H. I.
2. C, F
3. The simplified form of the expression is $3x^3-5$. Synthetic division can be used to simplify, because the expression's denominator is in the form $x-a$, where $a=-4$. Since synthetic division can be used, so can long division. Factoring and reducing can be used by factoring the expression's numerator as follows: $3x^4+12x^3-5x-20 = 3x^3(x+4)-5(x+4) = (3x^3-5)(x+4)$.

Develop Rational Expressions *(continued)*

Check for Understanding: Investigation 5

1. B

2. E, F

3. Changing $\frac{x+a}{x+b} \div \frac{x-c}{x-d}$ to multiplication $\frac{x+a}{x+b} \times \frac{x-d}{x-c}$ and multiplying gives $\frac{x^2+ax-dx-ad}{x^2+bx-cx-bc}$. Adding the fractions $\frac{x}{x-6}+\frac{2}{x+9}$ by first getting a common denominator $\frac{x(x+9)}{(x-6)(x+9)}+\frac{2(x-6)}{(x+9)(x-6)}$ gives $\frac{x^2+9x-2x-12}{x^2-6x+9x-54}$. Comparing the denominators $b=-6$, $c=-9$, $-bc=-54$. Comparing the numerators $a=9$, $b=-2$, $-ad+18$. It would appear that it is not possible because $ad=12$.

Practice Exercises

1. A, B, E

2. F
D
B

3. D

4. A, B, D
C, E, F

5. 1
−6
−6
9

6. A. −4
B. −28
C. −5
D. −20

7. A
C
B

8. B

9. A
C

10. D

Apply

What Does it Take to Be a Top Triathlete?

This problem allows students to demonstrate their ability to represent real-world relationships using rational expressions. Students solve rational equations and inequalities to answer the questions.

The Apply problems are rich performance tasks that may have a variety of open-ended solutions. One sample student response for this problem can be found in the Model Lesson.

15.2 Solve Rational Equations

Check for Understanding: Investigation 1

1. C
2.

Time for New Pipe to Drain Moat Alone (minutes)	Time for Both Pipes to Drain Moat Simultaneously (minutes)
25	16.7
40	22.2
55	26.2
70	29.2

3. Herb can stuff $\frac{80}{16} = 5$ envelopes per minute, and Eleanor can stuff $\frac{84}{12} = 7$ envelopes per minute. Therefore, it would take Herb $\frac{210}{5} = 42$ minutes to stuff 210 envelopes working alone, and it would take Eleanor $\frac{210}{7} = 30$ minutes, so $\frac{1}{42} + \frac{1}{30} = \frac{1}{t}$. Solving for t gives $t = 17\frac{1}{2}$ minutes. This makes sense because, together, Herb and Eleanor can stuff $5 + 7 = 12$ envelopes per minute, so they can stuff 210 envelopes in $\frac{210}{12} = 17\frac{1}{2}$ minutes.

Check for Understanding: Investigation 2

1. C
2. D, H, I
3. Both Ken and Lena are correct. With the denominators of the fractions written in factored form, the equation is $\frac{16}{(x-2)(x+8)} = \frac{10}{(x-2)(x+2)}$, so when Lena wrote the fractions with their least common denominator, she got $\frac{16(x+2)}{(x-2)(x+2)(x+8)} = \frac{10(x+8)}{(x-2)(x+2)(x+8)}$. Therefore, she got $16(x+2) = 10(x+8)$ when she set the numerators equal to each other. This is a linear equation with a solution of $x = 8$, and no extraneous solution.

Check for Understanding: Investigation 3

1. B
2. −3, −3, 0, 0, 9, 9
3. Conrad's conclusion is incorrect. Had he analyzed the denominator of the expression $\frac{x^2 - 2x - 8}{x - 4}$, Conrad would have found that when x is 4, the expression is undefined. His graph actually contains a hole at the point (4, 6), but it is so small, he can't see it. Since x cannot equal 4, the intervals in which x satisfies the inequality $\frac{x^2 - 2x - 8}{x - 4} > 0$ are $(-2, 4)$ and $(4, \infty)$.

Solve Rational Equations *(continued)*

Check for Understanding: Investigation 4

1. C

2.

Temperature (°F)	Hours for Ice to Melt
60	4
64	3.75
75	3.2
80	3
96	2.5

3. directly, directly, inversely

4. Suppose Ingrid lets t represent duration and r represent speed. If the distance of her trip is 250 miles, then she could write either $t = \frac{250}{r}$ or $r = \frac{250}{t}$. The shapes of the graphs of these two equations would be exactly the same. The only difference between the graphs would be the labels on the x- and y-axes. For the first equation, the x-axis would be labeled r and the y-axis t. For the second, the x-axis would be labeled t and the y-axis r.

Practice Exercises

1. A, C, E

2. C

3. 0, 4, 2, 12

4. B, D, E

5. D, C, D

6. A, B, E

7. 40

8. 1, 3, 9

9. B, F
A, D

10. E, B, C, B

Apply

How Big Should the Family Patio Be?

This problem allows students to demonstrate their ability to create a rational inequality to model a real-world problem. Students graph the inequality and interpret it to solve the problem. Explain to students that even though the calculator will most likely display more decimal places, that level of accuracy is not needed when pouring concrete. Students may make a table and estimate the solution, or they may use technology to create a graph and find the solution. The intersection of the graph of the rational equation and the line of $y = \frac{5}{12}$ is about 13.962, depending on whether 0.4, .42, or 0.4167 is used for the y value. So a range of values, 13.9 feet to 14.1 feet, can be used for the length, with appropriate other calculations noted.

The Apply problems are rich performance tasks that may have a variety of open-ended solutions. One sample student response for this problem can be found in the Model Lesson.

Notes

Notes

Notes